PLATO'S REVENGE

Plato's Revenge

The New Science
of the
Immaterial Genome

David Klinghoffer

Seattle   Discovery Institute Press   2025

Description

First there was the genetic revolution—the discovery that physical structures in the cell, including DNA and RNA, shape every organism. Now, says evolutionary biologist Richard Sternberg, we are overdue for another and more profound revolution. Recent findings reveal that genetic and even epigenetic sources alone cannot account for the rich dynamism of life—not even close. Some other informational source is required.

The idea was anticipated 2,400 years ago in Plato's *Timaeus*, and periodically revisited in the ensuing centuries. Sidelined by scientific materialism, it is now reasserting itself on the strength of cutting-edge molecular biology, higher mathematics, and commonsense reasoning. In *Plato's Revenge*, science writer David Klinghoffer takes Sternberg's profound explorations and weaves them into a lively and accessible account of a most remarkable realization: At every moment, we owe our lives to a genome that is more than matter, and to an informational source that is immaterial, transcomputational, and beyond space and time.

Library Cataloging Data

Plato's Revenge: The New Science of the Immaterial Genome
by David Klinghoffer
Cover design by Tri Widyatmaka.
146 pages, 6 x 9 inches
Library of Congress Control Number: 2025934900
ISBN: 978-1-63712-074-3 (paperback), 978-1-63712-076-7 (Kindle), 978-1-63712-075-0 (EPUB)
BISAC: SCI029000 SCIENCE / Life Sciences / Genetics & Genomics
BISAC: SCI072000 SCIENCE / Life Sciences / Developmental Biology
BISAC: SCI075000 SCIENCE / Philosophy & Social Aspects

Publisher Information

Discovery Institute Press, 208 Columbia Street, Seattle, WA 98104
Internet: discovery.press
Published in the United States of America on acid-free paper.
First Edition, April 2025

ADVANCE PRAISE

In 2004, for allowing impure thoughts about evolutionary theory to see the light of day, Richard Sternberg found himself targeted as a modern-day heretic. It was a shocking early instance of cancel culture. Despite what happened to him, Sternberg has never wanted his story to obscure the important idea he wanted aired: that the concept of the gene itself was deeply flawed and that hereditary memory is more complex, more dynamic, and more intelligent than the crude conception of the gene would allow.

David Klinghoffer's *Plato's Revenge* beautifully honors Sternberg's wish. He writes compellingly about Richard Sternberg himself and the controversy that engulfed him, revealing the shocking intolerance at the heart of Darwinism. More importantly, he frames Sternberg's story within the concept of the "immaterial genome," anticipating emerging challenges of epigenetics and gene expression that are undermining the gene as the "atom of heredity." In so doing, he brings to the fore the central, and largely unacknowledged dilemma of evolutionism: that how we think about evolution is as much philosophy as it is science.
—**J. Scott Turner**, Emeritus Professor of Biology, SUNY College of Environmental Science and Forestry, Syracuse, New York; author of *Purpose and Desire: What Makes Something "Alive" and Why Modern Darwinism Has Failed to Explain It*

Some thirty years after Francis Crick proclaimed the central dogma of molecular biology and its attendant materialist triumph, serious doubts began to emerge. This led another theoretical biologist, one of Rick Sternberg's early influences, Robert Rosen, to examine the question of *Life Itself* in rigorous detail. His conclusion regarding that reigning paradigm: "Something is missing, something big." Now, some thirty years after Rosen, in *Plato's Revenge*, David Klinghoffer masterfully chronicles Sternberg's decades-long search to fill the void articulated by Rosen, fulfilling the quest to which fate, or perhaps Providence itself, had assigned to his brilliant career. Klinghoffer's detailed yet accessible and engaging portrayal of such a demandingly wide-ranging topic leaves the reader breathless in anticipation of Sternberg's emerging synthesis of what almost certainly will prove to be one of the most momentous developments in the history of science.

—**Stephen Iacoboni, MD**, award-winning researcher, oncologist, and author of *The Undying Soul* and *Telos: The Scientific Basis for a Life of Purpose*

The mathematician-turned-philosopher Alfred North Whitehead famously said that all of Western philosophy consists of a "series of footnotes to Plato." Now what Whitehead said of philosophy may be applied to science. *Plato's Revenge* is about the teleologically ordered biological systems theory that Richard "Rick" Sternberg calls the immaterial genome. It is an ancient story that dates back to the atomists on the one hand and the teleologists on the other—Leucippus vs. Anaxagoras. The argument between reductionist evolutionists like Charles Darwin and design-oriented evolutionists like Alfred Russel Wallace harkens to these pre-Socratic sources, proving King Solomon's wise adage, "There is nothing new under the sun."

With Darwin the triumph of chance and necessity was considered complete. But one of the greatest teleological proponents of all history, Plato, now has his revenge as we find that design and purpose have won the day. The quantum physicist Werner Heisenberg understood this, saying, "I think that modern physics has definitely decided in favor of Plato. In fact these smallest units of matter are not physical objects in the ordinary sense; they are forms, ideas which can be expressed unambiguously only in mathematical language." Now Sternberg, as told eloquently by David Klinghoffer, expresses this in the language of life. This book unites the best elements of metaphysics with cutting-edge science to put the threadbare materialist reductionisms of the neo-Darwinists to shame.

—**Michael A. Flannery**, author of *Nature's Prophet: Alfred Russel Wallace and His Evolution from Natural Selection to Natural Theology* and *America's Forgotten Poet-Philosopher: The Thought of John Elof Boodin in His Time and Ours*

Science journalist David Klinghoffer's task is not an easy one: Introduce the general reader to a scientific hypothesis that is profound, potentially revolutionary, but hardly simple. Klinghoffer succeeds and, into the bargain, paints an engaging portrait of the scientist behind the idea, Richard Sternberg. As Klinghoffer explains, Sternberg has woven together the fields of biology, mathematics, and philosophy to argue that an organism's genome is not entirely contained in DNA. Moreover, the information representing a species' structures and processes is not confined to any physical molecule. Instead, an organism's architecture results from immaterial principles. Sternberg's arguments draw from the leading theorists who applied mathematics, such as category theory, to life, and his analysis demonstrates that the control center that directs an embryo to develop into an adult requires far more information than could be contained in the entire initial cell, let alone DNA. The control center must reside in a mathematical structure outside of time and space. Klinghoffer, following Sternberg, also traces scientists' understanding of the genome through history, illustrating that many leading biologists recognized the genome's immateriality.

—**Brian Miller**, PhD, Research Coordinator for the Center for Science and Culture, primary organizer of the Conference on Engineering in the Life Sciences, and contributor to *The Mystery of Life's Origin: The Continuing Controversy* and *Inference Review*

This combatively titled volume is in essence one long argument in favor of the philosophical underpinning pertaining to ideas of human existence bequeathed to us initially by the ancient philosopher Plato in his *Timaeus* some two and a half millennia ago. Lest any modern-day technocrats, exulting in our civilization's latest gadgets and gizmos, should think ancient philosophy superannuated and entirely dispensable, Klinghoffer reminds us of A. N. Whitehead's apt observation that modern thinkers have provided little but "a series of footnotes to Plato." Linking Plato with cutting-edge modern science research, the author makes much of the work of Richard Sternberg who, like Plato before him, posits an immaterial force beyond physical reality. Invoking up-to-date scientific findings he observes that neither DNA nor any other known epigenetic factors can account for us in the fullest sense. Beyond genetics and epigenetics there must lie some Platonic reality all its own.

Klinghoffer's arguments will inevitably affront those holding to materialist presuppositions, but glib objections will have to take into account the most outstanding desideratum of modern biological science, which the author locates in the fact that information sources beyond the material genome are required in order to answer that perennial question, What is life? This is an important contribution in that the author shows himself to be just as much at ease with modern scientific advances as he is with the still very timely thinking of Plato. The volume is to be highly recommended, and readers may be glad to hear that even the most rebarbative-sounding "hard science" is presented with an admirable lucidity.

—**Neil Thomas**, Reader Emeritus, University of Durham, author of *Taking Leave of Darwin*

There are more things than are dreamt of in our materialist philosophy—not just in heaven, but right here on earth. At a time when this is becoming newly, exhilaratingly apparent, David Klinghoffer's wonderfully readable book explores some breathtaking implications of the latest natural science in an area—genetics—that touches deeply on our origins, our characters, and perhaps our souls.

—**Spencer Klavan**, Associate Editor of the *Claremont Review of Books* and author of *Light of the Mind, Light of the World: Illuminating Science Through Faith*

Darwinian materialism fails to explain the biological information in DNA sequences. But that truth merely scratches the surface when it comes to explaining biological form. To understand organisms in all their complexity, argues Richard Sternberg, we must break completely with nineteenth-century materialism and reconsider the thought of ancient greats such as Plato and Aristotle. Sternberg's argument might seem daunting to the non-specialist, but David Klinghoffer does a masterful job of explaining Sternberg's revolutionary thought in a delightfully accessible way.

—**Jay Richards**, PhD, co-author of *The Privileged Planet: How Our Place in the Cosmos Is Designed for Discovery* and editor of *God and Evolution*

"Hold it up!" said Gandalf. "And look closely!"
As Frodo did so, he now saw fine lines, finer than the finest
pen-strokes, running along the ring, outside and inside: lines of
fire that seemed to form the letters of a flowing script. They shone
piercingly bright, and yet remote, as if out of a great depth.
"I cannot read the fiery letters," said Frodo in a quavering voice.
"No," said Gandalf, "but I can."

—J. R. R. TOLKIEN, *THE FELLOWSHIP OF THE RING*

To Günter Bechly, paleontologist and Platonist,
1963–2025

CONTENTS

INTRODUCTION

My view is not the view that most people have of intelligent design.
—RICHARD STERNBERG

WHEN I FIRST HEARD BIOLOGIST RICHARD STERNBERG DESCRIBE his immaterial genome hypothesis, reviving the thought of the Greek philosopher Plato in a modern and scientific context, another biologist on hand took in her breath. "If that's true," she said, "it changes everything." I felt similarly. The idea spooked me.

All familiar thinking about the genome assumes that it is, of course, purely material: the twisting strands of DNA and a few other physical structures in the cell. The proponents of intelligent design (ID) have, in large part, accepted this premise and argued according to its terms.

Sternberg goes further. He argues—sometimes from common-sensical and accessible evidence and sometimes from highly technical mathematical and biological realities—that the material resources of the physically instantiated portion of the genome are woefully inadequate to shape life from generation to generation. The conclusion still gives me a shiver: An immaterial source exists, in company with DNA and the other material sources of biological information. That source extends not only beyond us, but beyond physical reality.

I vividly recall the meeting. It was 2012, a time of great strain in my life. In a small conference room in Seattle, several of us, including scientists and non-scientists like me, gathered to listen to Sternberg sketch an argument he had been developing, rooted in his observation that there simply is not enough information physically in the cell—including the DNA and epigenetic (from the Greek, meaning *beyond genetic*) sources—to account for the development of an organism. According to him, this finding applied not only to the more complex organisms, such as whales and humans, but also to the relatively simple ones, such as yeast.

He also spoke of those who had influenced him—in particular, theoretical biologist Robert Rosen (1934–1998) and the men who had first devised the idea of a gene. As Sternberg explained, they saw it in terms that might not be material.

The nature of genetics and heredity is inherently of more intimate interest to some people than are many other scientific topics. The law of gravity says nothing about me as a person, except how fast I would fall if pushed off a tall building. Heredity promises to say much about who I am. As I am writing this, I have just put a tube of spit in the mail to the DNA ancestry company 23andMe. On the top of the test kit you get from Amazon is the message to the customer: "Welcome to You."

Siddhartha Mukherjee, who teaches medicine at Columbia University, begins his book *The Gene: An Intimate History* by detailing why the subject is painfully personal for him: Behind his narrative of scholarly discovery lies his father's family with its history of a mental illness, schizophrenia. In that family history he is himself implicated, as are any children he might have. "Madness," he writes, "has been among the Mukherjees for at least two generations." It is "buried, like toxic waste," in the genetic inheritance.[1]

In 2012, heredity was on my mind. That year, my birth mother, Harriet Lund, had come to live near me in the Seattle area, bringing emotional turbulence with her. She was suffering from dementia and, with it, episodes of rage and paranoia. It was at this time she told me that, in Los Angeles in 1965, my birth father, George Thomas, raped her. And this was how I was conceived.

She was Swedish-born, from a long line of Lutheran pastors. She was a social worker at the time, and George, a *Mayflower* descendant from Kansas, was her supervisor. I had first met her in 1993 and, charmed, wrote a book about her in relationship to my conversion to Orthodox Judaism. Yet she had not told me the crucial point about George Thomas until 2012, right around the time I first heard Sternberg's immaterial genome idea. Harriet sounded perfectly lucid when she said it: "Your father raped me! You're the son of a rapist!"

Later, after Harriet had already slipped away, present in her body but not in her mind and thus incapable of answering questions, a cousin of hers contacted me. The cousin revealed, with credible details, that Harriet had kept another secret as well: Harriet's own father—my grandfather, the Swedish filmmaker Oscar A. C. Lund—had sexually molested Harriet when she was a girl.

Before she became ill, Harriet had wanted to save me from the truth about my heredity. Only in the throes of dementia did she tell me the point about my birth father. Now I knew it all. As they put it at 23andMe, welcome to you. That is the first reason that Sternberg's discussion moved me.

There was something else about Sternberg that struck me. Given that he is a man decorated with two PhDs in biology—one in molecular (evolutionary) genetics and another in mathematical biology—and has held a scientific post at the Smithsonian National Museum, it's natural to expect him to have little interest in classical history. Here, surely, is a man oriented toward science, natural history, and the vanguard of discovery. But when you meet him in person, you quickly sense that a more complex description will be required. He is a man as interested in the history of science and philosophy as he is in the latest scientific evidence and ideas.

I confess that I find this very relatable. As a college student at Brown, studying Greek and Latin, I was narrowly diverted from an academic career in comparative literature. After graduation I was set to start in September in the Classics Department at Columbia to work towards a PhD. That summer, though, I was offered a job as assistant literary editor at William F. Buckley Jr.'s *National Review*. I deferred grad school for the coming academic year but then never went. The journalism virus had infected me. Yet I continued to find the Greeks, the picture of life that they offered, and their difficult language enchanting. More than this, I have remained fascinated by intellectual life as a form of archaeology, digging for insight and wisdom in ancient sources.

My children grew up hearing me say many times over that there are two kinds of people. There are those who see modern opinions as the product of an upward-driving, almost teleological evolutionary process, with a kind of natural selection picking out the very best concepts from what has come before: the more modern, the better. And then there are others like me who look around at contemporary existence, with its increasing surrender to mental illness as a philosophy of life, and conclude just the opposite. Jewish tradition calls this *yeridat ha' dorot*, the devolution of the generations. Human beings are not getting wiser. This is not only an axiom but is evident just from observing the world around you.

So I was naturally sympathetic to Sternberg, for though he grounds his argument for the immaterial genome in the latest discoveries of molecular biology, he is also by temperament sufficiently suspicious of novelty that he mined the history of philosophy and science to excavate intimations and intellectual forbears of his argument. In the process he turned up a line of thinkers and scientists from Plato to Rosen.

Recently, as I was reading about Sternberg, I came across the essay "Goethe and the Evolution of Science" by Craig Holdrege of the Nature Institute, an organization which seeks to revive the ideas of the German Enlightenment author Johann Wolfgang von Goethe about science and nature. As you will see, Goethe is among that line of great thinkers who influenced Sternberg. But I was struck by something Holdrege said. Sometimes people tell him, "Craig, we need to move beyond Goethe." He replies, "That has truth to it; but even more true is that in important ways we haven't even reached Goethe. That's why in going back to Goethe we can also move into the future."[2] Back to the future with Johann Wolfgang von Goethe.

In other ways, as well, Sternberg defies easy categories of description. He is a formal person, yet goes by "Rick." He has a precise manner of speaking and writing and the faintest hint of what sounds almost like a Southern accent. But, like me, he is from Los Angeles.

After his father died in 1969, he and his mother moved to Florida. There he attended a Catholic high school but was expelled just before graduation for persuading girls in chapel to let him pen on their legs, in all innocence, the legend "Rick was here."

A characteristic gesture of his is touching the tips of the fingers of one hand to the tips of the other, forming a bridge in five parts. When I first encountered him, he went by his legal name *Richard von Sternberg*, as he still does in professional publications.

My most recent in-person visit with him was in 2024 when he was in Redmond, Washington, for a meeting. A female colleague noticed that for a gathering of just three days, Sternberg, who had traveled from across the country, had brought two pairs of dress shoes. I would not have brought one pair, I thought at the time. He alternated between these from day to day. Because, after all, you would not want to be seen wearing the same shoes twice in a row. The same colleague was surprised one day when Sternberg wore jeans—though they were carefully pressed. On a subsequent occasion, when Sternberg was at the beach with his wife on a summer day, my colleague was startled to see him wearing shorts: "I thought he would be wearing long pants," she admitted.

Sternberg indulges in collecting art, and animals in terrariums and vivariums. His reading and learning styles involve highlighting phrases in various colors according to source and other categories. He has a thin mustache that makes him look as if he stepped out of the 1940s, perhaps playing a physician giving court testimony in a *noir* film.

He is given to certain formal expressions, like referring to admired scientists and other thinkers of the past as "a man such as James Clerk Maxwell," or "a man like William Bateson." He will occasionally curse. Otherwise, he is courtly and thoughtful, not only about science. When I saw him at the meeting I just mentioned, he referred to how, almost two decades before on another visit to Seattle, our wives had traded tips on swaddling babies. I did not remember that. While daunting in his intelligence and his encyclopedic memory, he is funny and humble. Standing at a whiteboard and mapping out a projected work on the immaterial genome, he summarizes his planned preface by writing on the board, "How I got to this state of conceptual insanity."

He is given to offering asides, often historical in nature. Endearingly, he is continually asking forgiveness for faults that don't exist and for things that need no apology. "I'm sorry for lingering on history," he will say in a lecture that greatly benefits from his stimulating historical digressions. And, "It's an interesting historical point from my perspective at least, maybe not from yours. I apologize for that if it's not." Or, "I've not done this justice at all, and I apologize for that." Or, again, "Well, I'll touch on that at the very end of this long-winded lecture. And I apologize for that."

What follows is not remotely as technical as a rigorous submersion into his argument would require. This is not an academic book. It aims only to relate the main lines of Sternberg's thesis. It looks forward to his much more technical volumes on the immaterial genome, which are in progress.

How, you might wonder, did a book like this about another man's ideas come to be? The pages that follow draw from or adapt interviews and conversations, recorded lectures and podcasts, and articles I wrote about him for the *Wall Street Journal*[3] and *National Review*,[4] as well as from other writings of his and of other people. Some conversations were in 2005, others in 2024 and 2025. Emily Sandico, working with Sternberg on a parallel project, offered insights from her own talks with him.

I extend my thanks to Dr. Sternberg for his openness, inspiration, and agreement to being portrayed, as well as for sharing some of his vast knowledge of biology and philosophy. He has read the manuscript and answered questions posed to me by the book's gifted editor, Jonathan Witt. But any errors of a scientific or other nature should be condemned as mine alone. And I apologize for that.

1. Scandal at the
Smithsonian

Richard Sternberg, thought criminal, first came to my attention in 2005, when I wrote an article about him in the *Wall Street Journal*.

Sternberg was then a research associate at the Smithsonian's National Museum of Natural History in Washington. He had been until recently the managing editor of a nominally independent journal published at the museum, *Proceedings of the Biological Society of Washington*, where he exercised final editorial authority. The August 2004 issue had included typical articles on taxonomical topics—for example, on a new species of hermit crab. It also included an article broader in scope, focused on the Cambrian explosion of animal phyla some 530 million years ago, and seeking a best explanation for this mysteriously rapid flowering of new animal body plans. The article: "The Origin of Biological Information and the Higher Taxonomic Categories," by Stephen Meyer.

And that was Sternberg's crime: to have edited and published a peer-reviewed article by philosopher of science Stephen Meyer.

Sternberg had met Meyer a couple of years earlier at a conference at Biola University in Southern California, attended by some ID proponents. (I had met Meyer in 1997, along with his Discovery Institute colleague Bruce Chapman.) Sternberg was speaking on a curious evolutionary phenomenon, bioluminescent organs in a type of fish, glowbellies. (Picture having a light that shines from within your intestines.) In the talk, Sternberg wondered, "What would be the causal entailment of evolving such an organ?" In simpler terms: How did that happen? As he argued, the organ appears to be irreducibly complex and therefore resistant to having emerged via gradual, step-by-step Darwinian evolution; and yet it arose not just once but convergently in a variety of teleosts.

"I was working on glowbellies when I was kicked out of the Smithsonian," says Sternberg. "I never got to complete that"—which, however, is getting ahead of our story.

Meyer was in the audience for Sternberg's Biola conference lecture and briefly introduced himself. They met again in 2003 in West Palm Beach, where they walked on the beach and talked about science, including biology and geology. That year, Meyer called Sternberg to tell him about the manuscript of an article concerning the Cambrian explosion that he had been working on, and asked Sternberg if he would be interested in considering it. Sternberg agreed to read the paper and found, he says, "nothing unpalatable. There was only one sentence that referred to intelligent design." It all sounded innocent enough.

That the Cambrian explosion is a problem for evolution was not Meyer's assessment alone. Far from it. As Meyer and paleontologist Günter Bechly later explained, "Most Cambrian experts agree that the majority of Cambrian animal phyla lack any putative fossil ancestors within the preceding Ediacaran biota. Thus the Cambrian explosion has been variously called"—here citing respectively *Time* magazine and Cambridge University paleontologist Simon Conway Morris—"'Evolution's Big Bang' and 'Darwin's Dilemma.'"[1] What's more, it is only one among a series of "biological Big Bang[s]," as biologist Eugene Koonin has put it.[2]

Where Meyer's article broke with scientific orthodoxy was in the explanation it identified as the best explanation. It patiently weighed mainstream explanatory options, all versions of modern evolutionary theory, and found them wanting for lack of a cause with the demonstrated capacity to generate so much new biological information and form so rapidly. What type of cause did Meyer point to as causally adequate? Intelligent design. The long essay was the first peer-reviewed article in a technical biology journal to do so.

According to the theory of intelligent design, certain features and phenomena of the natural world—such as the miniature machines and complex circuits within cells or the fine-tuning for life of the laws and constants of physics and chemistry—are better explained by an unspecified designing intelligence than by an undirected natural process such as random mutation and natural selection. The theory is compatible with at least one meaning of the mercurial term *evolution*—namely, change over time in the billions-year history of life on earth, including the overall ascent in geological time from less complex to more complex. The theory is not compatible with methodological materialism, since it posits an immaterial designing intelligence as the only causally adequate explanation for certain phenomena. Meyer identified the Cambrian explosion as one such phenomenon.

His review essay cited biologists and paleontologists critical of certain aspects of Darwinism—mainstream scientists at places like the University of Chicago, Yale, Cambridge, and Oxford. Meyer gathered the threads of their comments to make his own case. He pointed to the Cambrian explosion, when between 19 and 34 animal phyla (distinct body plans) sprang into existence. He argued that, relying on only Darwinian and other naturalistic mechanisms, there was not enough time for the necessary genetic information to be generated. Intelligent design, he suggested, offered a better explanation.

Meyer's 2004 paper cited then-recent work by biologists Gerd B. Müller and Stuart Newman, in which they argued that the "'origination of organismal form' in the distant past remains an unsolved problem."[3] This was a remarkable thing for evolutionary scientists to acknowledge.

Müller and Newman hoped that "epigenetic sources of morphological innovation" (sources beyond the gene, but presumably still material ones) would solve neo-Darwinism's problems. But meanwhile, they admitted it still was bereft of any "theory of the generative."

Meyer went on to survey other mainstream literature that hinted at or stated this outright. Given the evident inadequacy of the genetic sources, why were such material objects still looked to in many quarters as the sole information carriers? "Perhaps because the information carrying capacity of the gene could be so easily measured," wrote Meyer, "it has been easy to treat DNA, RNA, and proteins as the sole repositories of biological information."[4] However, haunting doubts about the "location of information in organisms" remained, a subject we will turn to in the next chapter.

If the scientific culture in and around the Smithsonian had been a healthy one, what you could have expected to follow from the publication of Meyer's paper was a civil, spirited, evidence-centered debate about his arguments and conclusions, either in a subsequent issue of the journal or in other journals. Short of this, such a discussion might have taken place online. This was not what followed. Instead, as we'll see, the knives came out.

To describe the reaction of Sternberg's higher-ups as unhappy is putting it mildly. Sternberg's editorship expired, as it was scheduled to anyway, but his future as a researcher was in jeopardy.

Apart from his work at the Smithsonian, Sternberg was employed as a scientist at the National Center for Biotechnology Information (NCBI), which is also part of the National Institutes of Health (NIH). The *Washington Times* reported at the time, "Mr. Sternberg... says he was 'called on the carpet' by his bosses at NIH after they were besieged by phone calls and e-mails from Smithsonian staffers, seeking his ouster."[5]

In a situation like that, having support from inside would be welcome. As it happens, among Sternberg's senior-most colleagues was Francis Collins, a prominent evangelical Christian and head of the National Human Genome Research Institute (NHGRI). The NHGRI is itself under the umbrella of the NIH. That made Collins a very powerful figure within the larger government organization where Sternberg worked. What was Collins's attitude in all this?

A little later, in 2006, Charles Colson of Prison Fellowship would take it on himself to invite Collins, Sternberg, and a few others for a dinner, says Sternberg, "to break bread and to talk." Collins himself, in an email to me, says, "I was aware of Sternberg's role in the publication in the Smithsonian journal, and as you know I was not a fan of the ID perspective on irreducible complexity, but I don't recall being particularly exercised about that publication." If you were some-one vulnerable like Sternberg, you wouldn't, of course, want to have an influential figure like Collins both "aware" of you and "not a fan" of something you had been controversially connected with.

By this point, Sternberg's involvement with the Meyer paper had been covered across national media, including the *Washington Post*, *Wall Street Journal*, and NPR. He was embattled. If Colson sought to create an atmosphere of Christian fellowship and support all around, it seems he didn't succeed. Sternberg recalls being intimidated by Collins's body language. Collins is a tall man—Sternberg estimates over six feet—while Sternberg is five foot ten. Collins, says Sternberg, walked up to within six inches of him, looked down at him, "and let me have it."

Collins, for his part, says, "I certainly don't recall a [one-on-one] conversation where I would have been intentionally intimidating to Sternberg—people who know me would say that would be very out of character for me." According to his recollection, says Collins, "the evening was warm and friendly."

Whatever the tone of the exchanges, at the gathering Collins insisted, according to Sternberg, that intelligent design illegitimately crosses the boundary between science and theology.

It's a curious complaint since Collins was, about the same time, publicly urging that, "for those looking to bring theologians and the scientists closer together, there is much in these recent discoveries of the origin of the universe to inspire mutual appreciation." Even more pointedly, "The Big Bang cries out for a divine explanation." The comments are from his book released that same year, *The Language of God: A Scientist Presents Evidence for Belief*.[6] Apparently Collins found it acceptable to cross that line in physics and cosmology, just not in biology, the domain dominated by pro-Darwin gatekeepers.

At the dinner, according to Sternberg, Colson pressed Collins on whether he thought God guided evolution, but Collins "wouldn't be pinned down." Sternberg recalls Collins's general response being along the lines of "Who am I, just this humble doctor, to speculate?" Colson seemed troubled by this.

Back at the Smithsonian, the treatment Sternberg received had turned openly hostile, so much so that he filed a complaint with the US Office of Special Counsel (OSC), saying he was subjected to discrimination based on his perceived religious beliefs. At the time, a museum spokesman confirmed to me that the OSC was investigating. Sternberg told me, "I'm spending my time trying to figure out how to salvage a scientific career."

What did the discriminatory treatment consist of? Part of it involved the baseless charge that he had only pretended to have Meyer's article peer-reviewed. In fact, whatever the article's ultimate merits, it was subjected to a thorough, blind peer review, the gold standard of academic science.

Not that sending it out for review saved Sternberg from infamy. Soon after the article appeared, Hans Sues—the museum's No. 2 senior scientist—denounced it to colleagues and sent a widely forwarded email calling it "unscientific garbage." The email was devoid of substantive evidence or argument against the article. "He had no leg to stand on. He was shooting from the hip," said Sternberg, mixing metaphors in his frustration. Sternberg recalls Sues as a "devout Christian, but of the Francis Collins type, a theistic evolutionist," and, moreover, a "creationist hunter"—someone who makes it his business to sniff out and go after suspected "creationists."

Sues wasn't the only one. The chairman of the zoology department, Jonathan Coddington, called Sternberg's supervisor. According to Sternberg's OSC complaint: "First, he asked whether Sternberg was a religious fundamentalist. She told him no. Coddington then asked if Sternberg was affiliated with or belonged to any religious organization.... He then asked... 'Is he a right-winger? What is his political affiliation?'"[7] The supervisor recounted the conversation to Sternberg, who also quoted her to me as observing, "There are Christians here, but they keep their heads down."

Worries about being perceived as "religious" spread at the museum. One curator told Sternberg about a gathering where he offered a Jewish prayer for a colleague about to retire. The curator fretted, "So now they're going to think that I'm a religious person, and that's not a good thing at the museum."

In October 2004, Coddington told Sternberg to give up his office and turn in his keys to the departmental floor, denying him access to the specimen collections he needed to do his job. Sternberg was also assigned to the close oversight of a curator with whom he'd had professional disagreements unrelated to evolution. "I'm going to be straightforward with you," said Coddington. "Yes, you are being singled out."

Sternberg successfully begged a friendly curator for alternative research space, and he continued to work at the museum. But many colleagues now ignored him when he greeted them in the hall, and his office sat empty as "unclaimed space." Old colleagues at other institutions refused to work with him on publication projects, citing the Meyer episode. The Biological Society of Washington released a vaguely ecclesiastical statement condemning the article and disavowing its association with it. It did not address its arguments but denied its orthodoxy, citing a resolution of the American Association for the Advancement of Science that defined ID as, by its very nature, unscientific.

The irony is worth noting: The scientists who complained that the article was unscientific responded to it not with reasoned, evidence-based arguments but with invective and stigma. And observe a glaring circularity: Critics of ID had long argued, and still do, that the theory was unscientific because it had not been put forward in a peer-reviewed scientific journal. Now that it had, they argued that it shouldn't have been because it's unscientific. They banished certain ideas from certain venues as if by holy writ and branded the proponents of those ideas heretics. And note: The heretic was Meyer, not Sternberg, who at the time was not an advocate of intelligent design. Sternberg was merely guilty by association.

One museum specialist chided Sternberg, saying, "I think you are a religiously motivated person, and you have dragged down the *Proceedings* because of your religiously motivated agenda." Definitely not. Sternberg told me he was a Catholic who attended Mass, but added, "I would call myself a believer with a lot of questions, about everything. I'm in the postmodern predicament." (Today, he asks me to note, he remains a Catholic with many questions, but with "more of a premodern mindset." As I mentioned earlier, he confounds easy descriptions.)

Intelligent design, in any event, is hardly a made-to-order prop for any particular religion. It counts among its proponents Jews, Christians, non-religious theists, Muslims, deists, and proponents of eastern religions. Darwinian materialism, by contrast, functions for many as an aggressive, quasi-religious faith without a deity beyond nature itself. The Sternberg case seems, in many ways, to be an instance of one religion persecuting a rival, demanding loyalty from anyone who enters one of its churches—like the National Museum of Natural History.

I should add that while the museum disputed Sternberg's account, the Office of Special Counsel did investigate and confirmed what he had told me, that the Smithsonian had created a "hostile work environment" for one of their colleagues "with the ultimate goal of forcing [him] out,"[8] merely because he had published a controversial idea in a biology journal. In August 2005, a lengthy and detailed letter from OSC attorney James McVay, addressed to Sternberg, summarized the government's findings, based largely on email traffic among top Smithsonian scientists. McVay described the attacks on Sternberg: "Retaliation came in many forms. It came in the form of attempts to change your working conditions… During the process you were personally investigated and your professional competence was attacked. Misinformation was disseminated throughout the SI [Smithsonian Institution] and to outside sources. The allegations against you were later determined to be false."[9]

Whether, in the end, Sternberg's legal rights were violated was never settled. Owing to what the OSC's McVay described as a "complicated jurisdictional puzzle," it turned out that because Sternberg was employed by the National Institutes of Health he was "effectively remove[d]… from the protections granted under the auspices of OSC."[10] Strangely, because Sternberg was formally employed by one government agency, the NIH, instead of another, the Smithsonian, he could not be protected by still another, the OSC. So the OSC investigation was closed.

When I contacted Sternberg in 2024 about my project of writing a book about his ideas, and asked for his agreement, he said he had one proviso: "I don't want to be portrayed as a victim." I assured him he would not be; and as the chapters that follow should make clear, there is much of the undaunted, trailblazing hero about his story. And yet, the Sternberg case was indeed an instance of retaliation by government employees. It was clear and troubling enough to be recounted in a motion picture, *Expelled: No Intelligence Allowed.* In more recent years we've seen other examples of how various figures in the government science sphere—including Francis Collins, who later headed the NIH—can stifle ideas they don't like. Sternberg's case foreshadowed things to come.

His case was important for another, more uplifting reason. The arguments and evidence in Stephen Meyer's paper were later greatly expanded in the form of three books, especially *Darwin's Doubt*, which became a *New York Times* bestseller, and most recently, *Return of the God Hypothesis*, endorsed by such distinguished scientists as Harvard Medical School's George Church and Nobel Prize-winning physicist Brian Josephson at Cambridge University.

And while that peer-reviewed article was the first to argue explicitly for intelligent design, it was far from the last. It's a talking point for evolutionists that in the past two decades, intelligent design has stalled. That is hardly the case. On the contrary, there are two very impressive measures of how much ID has advanced in that time. One is the latest update of the "Bibliography of Peer-Reviewed and Peer-Edited Scientific Publications Supporting the Theory of Intelligent Design," maintained by Discovery Institute's Center for Science and Culture. The full bibliography, with annotations, is itself the length of a book—186 pages in total, covering more than 170 papers as of this writing. (The actual number of pro-ID peer-reviewed papers is around 300, but to protect the authors, not all are listed publicly.)[11] That's not bad for such a young field and one facing down a dominant scientific paradigm whose proponents regard silencing opposition as just how business is done.

Another measure of ID's advance is the ID 3.0 Research Program. The program (following what proponents call ID 1.0 and ID 2.0) covers the years 2016 to the present and includes both pure and applied ID research. ID 3.0 is a much larger project than what can be revealed publicly, as many of these researchers would face the same censorship that Richard Sternberg did, and some papers can't be publicly acknowledged for the same reason. Among the more than twenty research projects under the ID 3.0 umbrella, with a more than $10 million budget since they were initiated, Sternberg himself is active in two. One is what is known as the "waiting times" problem (more on that further on). The other is the primary focus of the present book—the immaterial genome.

2. THE ELUSIVE GENE

RICHARD STERNBERG'S DARING IDEA CHALLENGES THE CONVEN-tional understanding of genes and the genome. But so as not to get ahead of ourselves, we should pause to lay out exactly what the conventional understanding is.

We live in an age where terms like "genes," "genetics," "good genes and bad genes," and "genetic engineering" are thrown around by even the most scientifically uninformed individuals. But ask a random sampling of people on the street to explain what exactly a gene is, and you are likely to get either blank faces or vague talk about DNA, with little idea how exactly the latter relates to genes.

The conventional scientific view of the gene can be found in popular online forums such as the MedlinePlus website, where the gene is defined as "the basic physical and functional unit of heredity," a unit "made up of DNA. Each chromosome contains many genes."[1] Siddhartha Mukherjee's book *The Gene* provides a more in-depth and technical account, though as with the definition at MedlinePlus, there is for Mukherjee little room for doubt that the genes are carried by a physical medium alone, some sort of "particle," a word that recurs in his telling. "There's just one molecule that carries our hereditary information and just one code," he informs us.[2]

The path to this view of the gene is long and winding. Pythagoras (circa 530 BC) and Aristotle located some heritable element in the semen, though the latter admitted that noncorporeal traits, often shared from parents to children, could not be conveyed that way. Mukherjee, like Sternberg, interprets Aristotle's word for "movement" to mean "instruction, or information—*code*, to use a modern formulation."[3]

Thinking had not really advanced much by the time of Charles Darwin (1809–1882) and Gregor Mendel (1822–1884). Darwin could only guess at the nature of inheritance. Anatomist Richard Owen (1804–1892) commented dryly, quoting Darwin against Darwin: "One's imagination must fill up very wide blanks."[4] Darwin's guess proposed "gemmules" as the particle of promise. The gemmules gathered information everywhere around the body and deposited it in the father's sperm and the mother's egg. Darwin called the idea "pangenesis," and admitted that it was a "rash and crude hypothesis." Mendel, the scientist known as the father of genetics, spoke of "elements" (close in meaning to what today we call "alleles"), whether dominant or recessive, "discrete pieces of information" inherited by the child from the parent.

Biologist August Weismann (1834–1914) disproved Darwin's gemmules[5] in 1883 by showing that mouse parents with their tails cut off consistently sired mice with tails. Botanist Hugo de Vries (1848–1935) argued in 1897 that "each trait was governed by a *single* particle of information." There is that word, *particle*, again. De Vries called the particle a "pangene." Biologist William Bateson (1861–1926), in 1905, contributed a new word, "genetics."

The science of genetics was ripe for abuse. Writes Mukherjee: "*If genes were, indeed, independent particles of information, then it should be possible to select, purify, and manipulate these particles independently from one another.* Genes for 'desirable' attributes might be selected or augmented, while undesirable genes might be eliminated from the gene pool."[6]

Such developments fortified the nascent eugenics movement, the inspiration of Darwin's cousin, Francis Galton (1822–1911), who could be found conducting his own scientific experiments, including the close observation of passing ladies, whom he graded on a scale from "attractive" downward to "indifferent" to "repellent." Besides beauty, he reasoned, high intelligence could also be inherited, notably in the greater families of England such as his own. In 1909, botanist

Wilhelm Johannsen (1857–1927) clipped the "pan" off "pangene" and gave us the gene. Mukherjee admits, "The word was created to mark a function; it was an abstraction."[7] In turn, the gene spun off two related concepts: the "genotype" (the information) forms the resulting "phenotype" (the organism).

At Columbia University with its Fly Room (the flies were fruit flies), working from 1905 to 1925, Thomas Hunt Morgan (1866–1945) and his students sought to "uncover the *physical* basis of heredity" (emphasis added): "The gene was not a 'purely theoretical unit.' It was a material *thing* that lived in a particular location, and a particular form, within a cell."[8] Morgan's student Hermann Muller (1890–1967) continued his work with flies, using radiation to stimulate genetic mutations. It worked. This meant, writes Mukherjee, that "genes had to be made of matter. Radiation, after all, is merely energy."[9]

This idea had profound implications beyond science. Although eugenic thinking might have found a way to marry itself to an immaterial view of the genome, it found a much easier match in the particulate view of the hereditary vehicle, which encouraged the attitude that inheritance could be manipulated as easily as one might manipulate different colored oil paints to achieve new colors. Hitler's Nazis with their "applied biology," inspired by America's own sorry record with eugenic sterilization (see Chapter 7 in John West's *Darwin Day in America*), recognized this.

Ironically, while Nazi Germany conducted genetic warfare, first against the disabled in Aktion T4, and later, in the Final Solution against the Jews, the Soviet Union took the opposite but also erroneous tack: denying the existence of genes however they might be understood. The party line of the 1930s, articulated by scientist Trofim Lysenko (1898–1976), held that "nothing about heredity was inherent at all. In nature, everything—*everyone*—was changeable. Genes were a mirage invented by the bourgeoisie."[10] They must be, or the Communist revolution would be seriously handicapped.

During World War II, physicist Erwin Schrödinger (1887–1961), in his lectures *What Is Life?*—delivered in Dublin in 1943 and published the following year—argued for the chemical nature of genes, a "variety of contents compressed into [some] miniature code." The same year biologist Oswald Avery (1877–1955) located these contents in DNA, "the material substance of the gene." The rest of the story, with James Watson and Francis Crick discovering the double helical structure of DNA and, with this, a way that it could be copied and passed on, does not need to be retold here. Mukherjee brings it full circle to Aristotle when he writes that at last the nature of the curiously named "movement" by which the embryo is formed, or informed, had been uncovered: "Embryogenesis could be reimagined as the gradual unfurling of gene regulation from a single-celled embryo."[11]

This is—or at least, has been for more than half a century—the ascendant view of the gene. But for the careful reader, notes of unease arise in this standard account. For instance, Mukherjee describes the research of biologist Sydney Brenner (1927–2019) at Cambridge University. Intent on understanding cells and their fates, Brenner worked with the tiny worm *Caenorhabditis elegans*. *C. elegans*, with about a thousand cells, is far from simple. (See the short video with ID proponent and philosopher of biology Paul Nelson, "How to Build a Worm."[12]) But it is a lot simpler than a human being with 37 trillion cells. Having mentioned that figure, Mukherjee notes something interesting: "A cell-fate map of humans would outstrip the computing powers of the most powerful computers." A cell-fate map, which Brenner sought for *C. elegans*, allows "*each cell* arising from the embryo [to] be counted and followed in time and space."[13] If such a map for a human being is beyond the power of our greatest computer, that raises an obvious problem. An embryo is not a computer. If all the information to "build a human" is contained in the embryo, in a material form, how and where is all that computing going on?

Sternberg and other scientists have been intrigued by such questions. The answer that Sternberg would eventually arrive at runs dramatically counter to the conventional physicalist view of the gene.

3. TRIPS TO THE
HERESY STORE

STERNBERG DESCRIBES HOW, AS A YOUNG PERSON IN THE SOUTH, he was alienated by religious fundamentalism and the young earth creationism that often goes with it. He studied Darwin's *The Origin of Species* and Richard Dawkins's *The Selfish Gene*. "By the age of twenty," he recalls, "I was an intellectually fulfilled atheist like Dawkins."[1] He pursued his interest in evolution as an undergrad at the University of South Carolina.

What knocked him off course was a bad lifelong habit: reading books. If you avoid unfamiliar ways of thinking, whether as a college student in your studies or as an adult following the news, sticking to sources of information that confirm what you want to believe, it's quite possible never to be in danger of changing your mind about anything serious. Casting your net more widely, though, is perilous. In doing so Sternberg eventually stumbled into seditious evolutionary works by Richard Goldschmidt, Hugo de Vries, Søren Løvtrup, and others, who criticized Darwinian theory not from a religious perspective but from a scientific one, from the evidence supplied by genetics and embryology.

"The 1980s were a time of upheaval in biology," Sternberg writes in an essay titled "How My Views on Evolution Evolved."[2] "So many revolutionary positions were being staked out in that decade—like pattern cladistics—that I lack the space to mention them." But geneticist Barbara McClintock's lecture upon receiving the Nobel Prize in

1983 made a major impact. She argued that certain "shocks," "unanticipated challenges," to a genome, to which the genome responds by reorganizing itself, can result in the formation of new species.[3] Her "jumping genes" led Sternberg to reject the prevailing view that the part of the genome that does not code for proteins was overwhelmingly junk. "Throwing caution to the wind," he says, "I decided to study the function of 'junk DNA.'" Such work, by him and others, would eventually lead to the discovery of widespread function among this "junk."[4] Sternberg was also struck by a 1988 paper in the journal *Science* by John Cairns et al., "on the evidence for Lamarckian-like directed (non-random) mutations in bacteria."

He spent considerable time hanging out with folks in the philosophy department. He calls them his "enablers." They "were only too happy to discuss theoretical biology with me over beers."

Sternberg's take on evolution was that of self-organization theory, which informed his first PhD, from Florida International University in molecular biology, where his studies included his "beloved junk DNA." He still saw the gene in conventional materialist and reductionist terms: "I had no problem accepting the premise that the development and morphology of the marine shrimps whose nucleotides I studied were specified by the genome," that is, by the material genome.[5]

The problem came with considering the shrimp as morphologic "wholes," not as mere static discharged by the DNA. Sternberg would later note, in the documentary series *Science Uprising*, the curious fact that the farmed shrimp you find in your shrimp cocktail have been around essentially unchanged in their body plan for 100 million years or more. Strangely, where you expect evolution, the *whole* persists.

Yet Sternberg still clung to certain traditional concepts about evolution, for, as he put it, "I knew from repeated trips to the heresy store that buying in bulk is expensive." Robert Rosen suggested Binghamton University (of the SUNY system) as a place where he could exercise intellectual freedom. It was good advice. In pursuing a second PhD there, "I began," says Sternberg, "to converse with men such as the late, great Ron Brady," as well as men such as Stanley Salthe, an agnostic biologist and "apostate Darwinian" at Binghamton.

Salthe "introduced me to the concept of a 'structural attractor,'" Sternberg explains, "an unchanging type of final cause that *informs* (my word) developmental processes."[6] The model, Sternberg realized, was recognizably Platonic even if, as Sternberg was careful to emphasize, Salthe himself would never go the Platonic route.

An allied view that Sternberg had also encountered in modern biology was that of the idealistic morphologists. "For years I had read about how bad these guys were," he says, "and how they were dangerous creationists." Their intellectual parent was Johann Wolfgang von Goethe (1749–1832), the great German poet, philosophical idealist, and scientist. Goethe, a "dangerous creationist"? Sternberg knew that was ridiculous misinformation peddled by Darwinists.

About this time he also first read the work of René Thom (1923–2002), the French mathematician, whose name occurs frequently in discussions of the immaterial genome.

The word "creationist," a Darwinist swear word, kept coming up as a term of abuse, employed to intimidate and keep doubters in line, as it would at the Smithsonian several years later. When Darwinists fling the term around, they undoubtedly hope their listeners will think of those who read the early chapters of Genesis as if they were a newspaper account and strain science in order to make it conform to that particular interpretation of scripture. Darwinists spray the term about with all the precision of a broken fire hydrant, as a tool to shame any and all potential heretics. But Sternberg was not trying to bend science to conform to anything. He was freely following his own intellectual lights. Of course, this put him in imminent danger. As he was finishing his second PhD in 1998, this one in Robert Rosen's field of theoretical biology, he was warned by a committee member that he'd better take care of himself; otherwise, he would "lapse into creationism."

Lehigh University biologist Michael Behe published *Darwin's Black Box* in 1996, and William Dembski published *The Design Inference* with Cambridge University Press in 1998, two central ID texts. Sternberg himself was hardly aware of intelligent design until 1999. Yet the peril of thinking his own thoughts was real.

His "rather hard structuralism"[7] of the time he saw more as an intellectual lark or pleasure cruise. "I found myself out on the conceptual ocean with no Darwinian shoreline in sight," he says, "and like the crew and passengers of the ill-fated S. S. Minnow on *Gilligan's Island*, my 'three-hour tour' landed me on an uncharted archipelago of ideas."[8]

It did not, however, maroon his career. In 2000 he landed a postdoctoral fellowship at the nation's beloved museum, studying crustaceans at the Smithsonian Institution's National Museum of Natural History. In 2001 he was offered and accepted a staff position at the National Institutes of Health, as well as an unpaid role editing a technical, peer-reviewed journal published from the Smithsonian, the *Proceedings of the Biological Society of Washington*.

Sternberg calls himself naïve for not seeing the trouble that publishing Stephen Meyer would cause him. Though he had given a presentation on "formal causation" to some ID scholars in 2002, in 2004 intelligent design still was largely unfamiliar to him. He knew design thinking had a distinguished lineage, going back even before Plato to the pre-Socratic Greek philosopher Anaxagoras. However, "I seriously believed the article would be mostly ignored," he remembers. Wrong.

"When the situation in the museum really became nasty," he says, "I could have used a stopwatch to mark the seconds from the start of a conversation about the Meyer article, to the tirades about Christians, 'fundies,' Republicans, George Bush, etc." The persecution that followed, he described as "surreal—like a David Lynch adaptation of a Kafka novel."[9] The rest of that story, what happened in 2004, has already been narrated here. Fast forward to 2024.

4. A Diffident
Revolutionary

In a meeting with colleagues at Discovery Institute in 2024, Richard Sternberg was sketching his thoughts on a whiteboard. What he was arguing for was revolutionary, but one would never have guessed it by his manner. He seemed to have taken advice from an icon of the American founding, Benjamin Franklin.

In his *Autobiography*, Franklin advocates for a modest way of setting your viewpoint before others, one he himself had employed to good effect and that involved "never using, when I advanced anything that may possibly be disputed, the words *certainly*, *undoubtedly*, or any others that give the air of positiveness to an opinion; but rather say, I conceive or apprehend a thing to be so and so; it appears to me, or *I should think it so or so*, for such and such reasons; or *I imagine it to be so*; or *it is so, if I am not mistaken*."[1] This "diffidence," he wrote, has the effect of making your opinion seem comelier and more acceptable than any more dogmatic approach.

In that presentation Sternberg was diffident in asserting his argument, a quality I will try to carry over into the summary of the argument below:

- Sternberg says that he conceives that the gene occupies "no *single* physical locus," that is, not a locus or place limited to DNA alone. This is already somewhat well known. He makes mention of regulatory RNAs, riboproteins, alternative splicing, the spliceosome, overlapping genes (don't worry about these or other technical terms for now), and the evidence that repetitive or junk DNA is not functionless but acts similarly to an operating system in a computer. DNA is like a library to be read, but who or what is the reader? Or it's like an art gallery, or a museum, but who is the curator?

• Moreover, beyond there being no single locus where the gene resides, it appears that the gene "has no *physical* locus" at all. Sternberg cites the Levinthal paradox, most familiar in its application to protein folding but expanded to the whole cell (in this case, yeast) by Peter Tompa and George D. Rose in a 2011 paper in *Protein Science*.[2] Moving from the single-celled yeast to the more complex form of a chicken, the paradox, Sternberg jots on the whiteboard, is that the "information output in a developed animal form exceeds the information present in the fertilized egg. Information transmission from egg to fully developed animal requires an external source of information to correct errors and an external channel of transmission." The information travels from elsewhere than the fertilized egg, and it does so by no "channel" that existed in the egg. He summarizes: "The information required to build the fully developed animal (the chicken) will not fit in the fertilized egg." *If I am not mistaken*, this means the information comes from a place outside the animal, and that place could not be a material place, since the channel by which it travels is not material. In connection with these matters, he notes the "problem of specifying future states while course correcting," Claude Shannon's 10th theorem,[3] an 1875 entry on the atom by James Clerk Maxwell in the *Encyclopaedia Britannica* (9th edition), Walter Elsasser in *The Physical Basis of Biology*, and Robert Rosen's 1991 book *Life Itself* and a 1999 follow-up, *Essays on Life Itself*. To his argument, Sternberg adds many modern scientific papers and discoveries, as we will see.

• In conclusion, *he apprehends* that the genome is immaterial in nature.

The basic and important question, of course, is whether his conclusion is true. Closely tied to that is the question of how well supported it is by the evidence in biology and related fields. Less basic and less important but still of interest is whether Sternberg, in making such an argument, has fallen in with that infamous outgroup, the proponents of intelligent design.

Sternberg explains that "I look at the whole ID issue from the standpoint of neo-Pythagorean Neoplatonism." He adds that such a response is "apparently often seen as an evasion by means of high-sounding metaphysical labels or an attempt at obfuscation."[4] Yet it is less strange for a scientist to speak this way than it might seem. As this book was nearing completion, biologist Michael Levin, with dual appointments at Tufts and Harvard and not an ID advocate, published a paper arguing, in his own words, "for a Pythagorean or radical Platonist view in which some of the causal input into mind and life originates outside the physical world."[5]

What does neo-Pythagorean mean? It means, says Sternberg, that "the universe—including every object in it and all relations between and among those objects—has its basis in logical-mathematical structures."[6]

He cites decorative seashells, the exquisite patterns of butterfly wings, the shapes of flowers and of mollusk shells. I think of my pet garden snail, Sancho, collected from outside my building after a rain one night. The house he carries on his back features a gorgeous colored pattern, chartreuse with an ironic dark brown racing stripe. Sternberg perceives a "transcendent form" behind such phenomena.

As for the source of the form, he thinks his Platonism and structuralism "enable one to remain agnostic," but do not preclude "holding that the structure emanated from *Nous* (mind) or *Logos* (intellect)." His own view was held by such "pagan 'saints'" of the ancient world as the Neoplatonist Plotinus and the "venerable and virginal Hypatia." "I hold with Plotinus on down to Proclus that the Realm of the Forms proceeds from the One, which is beyond all Being," he tells me in an email. "That said, this Realm is 'in' the *Nous*, which emanates from the One. In turn, all that is intelligible in the realm of coming-to-be emanates from the *Nous* and, hence, from the Forms." (Don't worry if you aren't entirely following this. More explanations and clarifications are to come.)

He distinguishes this idea, understandably, from what he describes as the "arid deism" and the "most unattractive and legalistic moralism" of a Francis Collins, a figure presuming to know what God would have done and rejecting the design hypothesis on that basis.[7] In an argument Collins offered to Sternberg on one occasion, the former's theistic evolutionary logic went like this: If I were God, I wouldn't have created dead pseudogenes (a form of "junk DNA" to which we'll return for a definition in Chapter 10). So God didn't either.

Sternberg rejects attempts to read God's mind, saying, "I for one have no knowledge concerning how God would or would not have created, or concerning His specific artistic tastes, nor does any other scientist."[8] If Sternberg feels closer to the pagan saints than to Francis Collins, I can identify. It is an unsure enterprise to confidently declare what God would or would not do in any context, including natural history. And even if one grants the assumption that a wise designer would be disinclined to create dead pseudogenes, there is the additional problem of their not being dead, as it turns out.[9]

That is a brief profile of Sternberg's thinking. I don't care for the word "unpack" as a synonym for "explain," as if the head of a person with a complex, elusive, and beautiful idea in it were like an overstuffed suitcase at the end of a long day of travel, when you need to separate your underwear from your folded shirts and your toiletries bag, to be placed in the appropriate drawers in the hotel room. But to be appreciated, Sternberg's thinking does need to be separated, unfolded, carefully pressed, and placed on coat hangers if appropriate. In a two-part class on the immaterial genome, for Discovery Institute's Summer Seminar on Intelligent Design in the Natural Sciences, he went deeper into the historical background and into the idea itself. I want to present that to you. Let's unpack it.

5. A Process,
Not a Particle

Sternberg begins his lecture[1] on a formal note, like a college professor from the 1950s: "Well, hello, ladies and gentlemen." His subject is the *real* history of genetic thinking, not the conventional one we reviewed earlier. His theme is that the gene is "not a particle, but a process,... a multilevel mediator of information that currently lacks a material description."

Sternberg starts by noting the disappointment of the Human Genome Project. When the genome was sequenced—"purportedly sequenced,"[2] he says—by two groups, including Francis Collins's National Human Genome Research Institute, part of the National Institutes of Health (NIH), the expectation was that scientists would now be able to read out the genetic book of life. Having read all those nucleotides, those A's, T's, G's, and C's that constitute the instructions in DNA, it would be like a recipe from a cookbook. Assemble noodles, butter, onions, frozen peas, canned tuna, cream of mushroom soup. Soon enough you have a tuna casserole. We were promised "personalized medicine," where your propensity for a certain disease could be read out from your particular genetic recipe, and then seamlessly addressed by you and your physician. Sternberg was then, in early 2001, on his way to becoming a new staff scientist within the NIH and assumed, like his colleagues, that "we'll be able to read... the blueprint for making a human being, and it should be more or less self-explanatory."

It was not self-explanatory. "The problem," says Sternberg, "is that as soon as you had millions to billions of nucleotides assembled, no one knew really how to decipher it."

Evolutionists[3] wrote off the vast majority of the genome as evolutionary detritus, junk DNA left over from evolution's blind trial-and-error process of evolving new biological forms. Then in 2012 came the ENCODE (Encyclopedia of DNA Elements) project. Its results were unexpected to most evolutionists, but not unexpected to proponents of intelligent design. ENCODE showed that some 80 percent of the genome is transcribed from DNA to RNA, or otherwise biochemically active, implying genome-wide function. On the junkiness of DNA, Darwinists were shown to be wrong, but leading spokesmen simply absorbed the news and changed their tune without an acknowledgment or an apology. (It is hard for humans to admit they are mistaken.)

Atheist biologist Richard Dawkins is a case in point. In his 2009 book *The Greatest Show on Earth*, he wrote, "The greater part (95 per cent in the case of humans) of the genome might as well not be there, for all the difference it makes."[4] Got that? Supposedly, 95 percent of the human genome is evolutionary garbage. And that would make sense if Darwinism were correct. But three years later, after ENCODE had indicated widespread function in the genome, putting the "junk" thesis on its heels, Dawkins turned his earlier contention on its head. In a conversation with Rabbi Jonathan Sacks, he now argued that widespread function was just what Darwinian evolution would expect: "It's exactly what a Darwinist would hope for, is to find usefulness in the living world."[5] Give Darwinism points for flexibility.

Francis Collins, whom we've met already, a theistic Darwinist and evangelical Christian, was more transparent about the changed thinking. Like Dawkins he was on board the junk-DNA bandwagon, asserting in his 2006 book, *The Language of God*, that up to half of DNA is garbage, "made up of... genetic flotsam and jetsam."[6] But nine years later, at a professional conference, he conceded that "we don't use that term [junk DNA] anymore. It was pretty much a case of hubris to imagine that we could dispense with any part of the genome—as if we knew enough to say it wasn't functional."[7] Good point. But by 2024, in a new book, *The Road to Wisdom*, written for the same popular and mainly evangelical Christian audience for which he wrote *The Language of* God, he simply dropped the junk DNA argument in silence, without admitting the error in his earlier book. Thus, if readers of *The Language of God* missed his concession to professional colleagues, they missed the concession entirely.

Yet what was all that former "junk" doing? Why was it being transcribed? The answer, says Sternberg, is that the nature of biological information is "far, far more complicated than it was assumed to be," and "far, far more intricate than originally thought." The problem lay with previous assumptions that heredity was a matter of tiny material units, like "small beads on a string," that spelled out how to generate an animal.

Aristotle understood the falsity of this, and "used logic, not microscopes, to refute" the materialism of earlier and contemporary philosophers. Sternberg is impressed by the "many points that Aristotle got right on embryology." Against the idea of pangenesis, of inheritance by particles, Aristotle gave an analogy to writing.[8] Sternberg explains, "What takes those particles and builds them up into an embryo? The analogy is to letters, composing syllables, composing words and phrases and sentences. What's generating that language? Is ink composing it? No, there has to be a source of agency."[9] There is an end goal, a disembodied *telos*, that "attracts," says Sternberg, the embryo's development. "It's almost the inverse of the modern conception," and yet one that he thinks is correct.

The idea that inheritance is "particulate"—transmitted piecemeal through what we've called in modern times the material genome—has a long history. Sternberg traces it from ancient Greek thinkers down to the French philosopher Pierre Louis Maupertuis (1698–1759), who anticipated Darwin. On this view, says Sternberg, "there are these elementary particles that are passed on from parent to offspring, and indeed they're mutable. And if they're mutable, and they specify the adult, then you could imagine that you'd have a number of new species appearing all the time." Particulate inheritance seemed to help explain a mystery like insect metamorphosis, as in the caterpillar which fully dissolves into a goo within the chrysalis only to reassemble as the beautiful butterfly that emerges. Charles Bonnet, the Swiss scientist (1720–1793), was "shocked" by this until he theorized that it was the hereditary particles that were behind the dissolution of most of the larval (caterpillar) parts and their reassembly during pupariation and pupation into the adult butterfly/moth form. "What really stays true are these particles," explains Sternberg in continuing to limn the particulate theory, "and these particles can be understood in materialistic and mechanistic terms."

Left aside from any such materialist conception was a compelling notion of what was directing the particles to perform these wonders, as if the tuna casserole would read its own recipe and assemble itself.

In 1864 the English biologist Herbert Spencer (1820–1903) proposed that the gametes—in humans, the egg and the sperm—carry "physiological units." This way of thinking was important to Charles Darwin's theory of natural selection with, in Darwin's 1867 articulation of it, the mythical gemmules performing the role of hereditary particles. As Sternberg says, most people still think in these terms: "Oh, I have my mother's eyes. Oh, I'm bald because of some genes that were passed down from some distant ancestor or not-so-distant ancestor." Darwin, philosophically innocent, had no notion that Aristotle and philosophers before him had already debated the matter.

Darwin's German follower Ernst Haeckel (1834–1919) replaced gemmules with "plastidules," another imaginary material entity. Haeckel, a gifted if not always honest illustrator, even sketched the plastidules. "He provides this beautiful diagram of these particles that grow, take in nutrients, and then divide," comments Sternberg. "And these particles are the bearers of various traits that go to make up the organism."

That is all very remarkable for particles which do not exist. There was something else remarkable about them: Despite being fully material, they were said to have "souls" enabling their performances, their ability to "self-develop."

These ideas were in the air at the time, and were, says Sternberg, a necessary entailment of Darwinian evolution. A materialist theory requires a physical means by which inheritance occurs. Once Darwin had proposed his theory, "then all of a sudden you need to have some materialistic, mechanistic basis for that." A number of scientists came forward with the required theories, which, as Sternberg points out, are like throwbacks to the Greek philosophers that Aristotle contended with. Not that these post-Darwinian scientists should be mocked for it. "One of the most notable figures of the time," says Sternberg, "and an individual for whom I have a great respect is August Weismann." During the early to mid-1880s, the German biologist originated the idea of a "germ-plasm," to explain inheritance from generation to generation.

Darwin, recall, had allowed for his gemmules to be modified by external factors, a Lamarckian-like postulate that Weismann doubted. "Weismann's mouse experiments were designed to see whether the 'germ-plasm' could be altered by such a means," Sternberg comments. "By finding that 'cut tails' were not passed on from parent to offspring, and by holding that germ-plasm consists of directive chromatin particles, Weismann was able to propose that the latter internally guide development and do so in a way that cannot be modified by the organism's environment. So, by closing the door to Darwin's allowance for inheritance of acquired characteristics, he opened the other door that led to neo-Darwinism."

"He was a fine zoologist," Sternberg adds, "and he had some inkling of what would be needed for the particles to perform the work they were supposed to do." Weismann began preparing the ground for what would come to be called neo-Darwinism, and indeed, he laid "the foundation for pretty much the current view, that chromosomes are the end-all and be-all" of heredity.

Sternberg challenges that view. To be clear, he accepts things like chromosomes carrying DNA which encodes proteins, and he understands that these are all very important. That much is not controversial. He does, however, think they are far from the whole show, and even doubts that purely material sources of information in the operative cell, genetic plus epigenetic, suffice. He questions whether that is all there is to it. "And," he says, "that scares a lot of people because we have a commitment to a notion of a closed universe"—without dimensions or realities that transcend our own—"and you find this regardless of where you are on the spectrum."

Paradoxically, one finds a version of this prejudice even among some intelligent design proponents, he suggests.

To be sure, intelligent design as formulated by leading figures in the ID community, especially if they see the combination of biology and cosmology as pointing to a "God hypothesis," is not materialist. The biological "big bangs" starting with the Cambrian explosion, those geologically sudden infusions of information into the biosphere, cannot have originated from the material world, unless one wants to invoke ingenious aliens—and leading ID theorists, including Meyer and Dembski, most definitely do not. Moreover, one cannot even make a desperate reach for aliens in the case of the fine-tuning of the laws and constants of nature at the Big Bang, since the Big Bang precedes all such physical existence. Sternberg and these ID proponents agree that more than mindless matter is required, that an infusion of information from an immaterial source is needed.

Yet, according to Sternberg, "most people who would be friendly to ID, or would say, 'Oh yes, I fall into the ID camp,' are in one way just as materialistic as any run-of-the-mill Darwinian, because they have this notion that whatever was passed on," the genetic inheritance, "was given at some distant point in the past." At the very least, they tend to be carefully agnostic about its having been partially informed by a cause from outside the universe during embryological development.

Sternberg's view is intelligent design taken to a new level. Call it (with apologies to the ubiquitous online merchant) Intelligent Design Prime. According to this view, we can securely infer that purposive design occurred not just once upon a time, but also in the here and now—a purposive infusion of form as an ongoing activity.

6. The Fathers of Genes

Here Sternberg turns a decisive corner in his lecture. He explains that the counterpoint to materialist genetics traces down from those "pagan saints" he admires and emerges again in relatively modern thinkers, including Gregor Mendel.

"Often in textbooks," says Sternberg, "you'll see the comment that the father of particulate inheritance is Gregor Mendel." But this is different from how Mendel saw his own contribution to science. In his writing about pea plants, at least to judge from the language he used in his 1866 paper, "Experiments in Plant Hybridization,"[1] he was not seeking to propose a physical mechanism: "What Mendel says is that he noticed some statistical regularities between parents and offspring. They can be put in a notation, and that's exactly what he did." The father of genetics, however, did not understand himself as having discovered "genes," in our conventional sense of "particulate units."

Instead, what he believed he had discovered, says Sternberg, is a "formal law," a law for the assortment and segregation of characters— that is, the observable characteristics of the plants he studied. For example, is the plant tall or short? "He's not necessarily talking about things"—material things—"that are passed on" from generation to generation. Sternberg shows this from a close reading of Mendel's language. In his 1866 paper, the word for "character" appears 157 times, the word for "elements" ten times. He speaks of a *"Beschaffenheit,"* or "inner constitution," again ten times.[2] All of this has the feeling of abstractness.

As other examples, Sternberg cites "such men" as biologist William Bateson and physicist James Clerk Maxwell.[3] Bateson, in *Materials for the Study of Variation* (1894), "goes through the arguments that were being played around with at the time, and he just says they're bankrupt and you have to start afresh." He was clear about this in 1902: "We have no warrant for regarding any hereditary character as depending exclusively on a material substance for its transmission."[4]

A prejudice favoring the materialist view of the genome (those marbles, those beads) was introduced by technology: You could see chromosomes under a microscope. Says Sternberg, "And chromosomes are called chromosomes because they're what? Well, in Greek, *chromosoma*, colored bodies, because you could stain them. So you said, 'Aha. I've found the material basis for heredity.' And this is interesting because the men who gave us some key terms in genetics didn't buy any of this." In contrast, an immaterial genome cannot be observed, only inferred from its effects. Thus, the idea will always have to face down the physicalist prejudice.

The key terms contributed by botanist Wilhelm Johannsen, for example, include "gene," "genotype," and "phenotype." Johannsen was careful to distinguish causation from correlation. "He wasn't convinced that simply because you could see something being passed on from a mother cell to daughter cells, that one had identified the causes of why it was being passed on," says Sternberg. "He was very skeptical of that, for purely logical reasons and also because he had read a lot of Aristotle." He knew these debates had been going on for millennia and that Aristotle had rejected particulate inheritance on logical grounds.[5] Pangenesis, which attributes heredity to particles that make their way into the offspring, was not Darwin's invention but was an idea that had been kicked around since Hippocrates and Democritus. Wilhelm Johannsen wrote extensively on this philosophical background.

Yet didn't he give us the very word "gene" itself? Yes, but the term was not intended to convey any hypothesis about the nature of the gene. Says Sternberg, "It was just supposed to be a label that one could use to track how traits appear and reappear in generations." It is a "placeholder," "and just because you can put a label on [a gene] doesn't mean that it's a particular physiological unit." As for the word "genotype," Johannsen intended that again as an abstract concept, a mathematical one, reflecting his skepticism about the thinking that had come before him, since Darwin and well before that. It was a new word for a new way of trying to conceptualize heredity.

Sternberg offers an analogy. Johannsen saw the pangenesis (or particulate) view "as trying to explain a locomotive by saying that you have miniature horses in there that are actually doing all the work. The locomotive would be the hereditary process, and the little horses would be the organoids or vital particles. And so he was like William Bateson, like Alfred North Whitehead, like many others, and I would say other geneticists much later, who were very skeptical that chromosomes are the end-all and be-all."

This skeptical attitude was prominent into the twentieth century. Thomas Hunt Morgan in his Fly Room at Columbia shared it with Bateson and Johannsen. In 1915, Morgan and three co-authors published their book, *The Mechanism of Mendelian Heredity*. In the preface, they allude to the divide in opinion among scientists about the nature of inheritance:

> Exception may perhaps be taken to the emphasis we have laid on the chromosomes as the material basis of inheritance…. But it should not pass unnoticed that *even if the chromosome theory be denied*, there is no result dealt with in the following pages that may not be treated independently of the chromosomes; for *we have made no assumption* concerning heredity that cannot also be made abstractly without the chromosomes as bearers of the postulated hereditary factors.[6]

Sternberg points to the references I have placed in italics above as evidence of exception being taken, and of the material basis of inheritance being called into question. The apparent skepticism could be because an immaterial basis of heredity remained a live idea, for logical reasons. And this being so, Morgan et al. were careful to make no assumption about the chromosomes as "bearers" of the genetic inheritance. Certainly a strictly mathematical analysis does not rule out the immaterial hypothesis. Sternberg paraphrases Morgan and his colleagues: "You don't have to buy the materialistic hypothesis because this all holds up mathematically, and we can just deal with the mathematics." Morgan himself, says Sternberg, seems to have remained "very skeptical" on the question of the gene's materialistic nature right to the end of his life in 1945.

Sternberg speculates on a reason underlying the rift between the materialists and skeptics. It has to do with the legacy of "the V word," as he calls it: vitalism, the idea that a "vital force" operates in life, a force that may be either supernatural or fully natural. The idea fell out of favor not because it failed scientific tests but because of the supernatural taint. Provocatively, science writer Daniel Witt has compared it to intelligent design, in that it offends, not because it can't explain the evidence, but because the gatekeepers don't like *the way* it explains.[7] Skepticism about a fully material basis for heredity also offends, and for much the same reasons. And so a word needs to be said about vitalism, that "dreaded term."

"I'll simply say this," Sternberg comments. "I do not propose vitalism. I'm not a proponent of vitalism. But one has to be cautious about the term 'discredited' that's applied to it." The word can mean one of two things. "One is where someone says, 'Well, we put an idea to the test, and it was found wanting.' But there's another kind of discreditation, and that is where a community simply says, 'We don't like the notion, and therefore if you come up with that argument, we're going to say the notion has been 'discredited' and we're going to shut you down.'" I believe he is recalling here his own history with the Smithsonian Institution, as well as the broader debate about intelligent design. He adds, "If you don't think that happens"—people being discredited in the sense of being arbitrarily canceled—"well, I don't know where you've been for the past few years."

7. Life Itself

THE DISCUSSION OF HEREDITY TOOK A NEW TURN WHEN PHYSICISTS considered biology and began applying the insights of quantum theory. They conducted thought experiments, applied logic, and found something lacking in the materialist model.

We have by this point advanced closer to the middle of the twentieth century. John von Neumann (1903–1957), who was crucial to the development of the computer, noted that the self-replicating gene, if considered under the heading of code, must be purposive because a computer program always reflects the programmer's purpose. Others seemed to go even further.

Erwin Schrödinger may be best known for his thought experiment in which no cat was harmed: the famed Schrödinger's cat, confined in a box, and both dead and alive at the same time. In his 1943 lectures, he says that heredity, dependent on a "code-script" recorded in an "aperiodic crystal," presents a paradox: "The chromosome structures are at the same time instrumental in bringing about the development they foreshadow. They are law-code and executive power—or, to use another simile, they are architect's plan and builder's code—in one."[1]

Sternberg sees in Schrödinger's formulation not a paradox but an echo of Aristotle: "Whatever the gene is, it has to combine—and these are my words, not his words, but he says effectively the same thing—that it has to combine all four Aristotelian causes: the material, the efficient, the formal, and the final."

The aperiodic crystal is DNA, but says Sternberg, Schrödinger's notion of heredity is more complex than is often imagined: The "code-script idea" could not be taken too literally and required "going over and above the laws of physics as they were known in the 1940s." Niels Bohr and Werner Heisenberg felt similarly that life could not be explained under the then-known physical laws. Still another physicist, Nicolas Rashevsky, felt that "if you want DNA in the context of the cell to be the whole show, you run into some serious problems," notably "the fact that the whole system [that is, the cell] divides."

If DNA isn't the "whole show," then what other forces in heredity were at play? Other physicists, meanwhile, worried that such thinking went too far, says Sternberg, that "mystical properties" were being attributed to the gene, and "it was suddenly becoming something that you could not explain biochemically."

Among those noticing the problem of *Life Itself*, as he put it in the title of his 1991 book for Columbia University Press, was Sternberg's intellectual guiding star, the biologist and biophysicist Robert Rosen. He "was the first one to apply category theory to the question of what are we talking about when we mention genotype and phenotype," says Sternberg. "What is really being meant? And by 1958, the very year that Francis Crick codified the sequence hypothesis in the central dogma, Rosen works out a very simple-looking—I would say set theoretic; it's more set theoretic than anything else, but it's in the domain of category theory—explanation for what these things are."

A thorough explanation of set theory is beyond the scope of this little book, but the *Stanford Encyclopedia of Philosophy* defines it as the "mathematical theory of well-determined collections, called *sets*, of objects that are called *members*, or *elements*, of the set."[2] It studies the infinite, and explores the curious fact that some infinities are greater than others. This is easily demonstrated. The mathematician Georg Cantor in 1873 noted that the points on a line cannot be counted, either in natural numbers (the positive integers—that is, the counting numbers) or in real numbers (which include all natural numbers as well as negative integers, fractions made from integers, and irrational numbers such as pi). Both sets are infinite and yet it is obvious, explains the *Stanford Encyclopedia*, that "there are more real numbers than there are natural numbers." Not all infinite sets are created equal.

Rosen found this relevant to understanding genes. Says Sternberg, "Whatever is going on with these abstract entities that are supposedly capable of self-duplication and self-multiplication, and all these other properties of being, if you will, self-predicating, this is something that you're just not going to get from your average chemical or physical system." The gene, in other words, transcends the physical: a very Platonic insight. Through the 1960s, "you suddenly have this concern being raised by logicians, mathematicians, and physicists that 'I'm not sure you've got the whole story here.'"

In heredity, it seemed there was something going on beyond the chemical or the physical. Rosen took aim at reductionism, the idea that "every perceptual quality can, and must, be expressible in numerical terms."[3] We should note that we have arrived here at a dogmatism or prejudice that is related to but separate from physicalism. One need not be a physicalist to argue that everything is reducible to numbers, and indeed that position doesn't even lend itself particularly well to physicalism. After all, the Pythagoreans saw numbers at the foundation of everything, and they were far from physicalists.

Reductionism in Rosen's sense holds that all experience, life itself, is reducible to numbers or the mathematical. That outlook lends support to the view that everything in science, including biology, is "ultimately a special case of physics," given that physics informs chemistry and biology at a fundamental level, and given that physics is the most mathematical of the natural sciences. This form of reductionism is related to the "formalist position" that reduced everything to "word processing or symbol manipulation," "meaningless symbols pushed around by definite rules of manipulation." This was the reductionism of conventional thinking about the genome.

It implied that life ultimately could be mimicked by a machine or automaton, an idea advanced by René Descartes (1596–1650) centuries earlier.[4] Rosen explains:

> Apparently, Descartes in his youth had encountered some realistic hydraulic automata, and these had made a great impression on him; he never forgot them. Much later, under the exigencies of the philosophical system he was developing, he proceeded to

turn the relation between these automata, and the organisms they were simulating, upside down. What he observed was simply that automata, under the appropriate conditions, can sometimes appear lifelike. What he concluded was, rather, that *life itself was automaton-like.* Thus was born the machine metaphor, perhaps the major conceptual force in biology, even today.[5]

According to the metaphor, biology is the study of machines. This is the same discussion our culture is having today about artificial intelligence and its supposed ability to do all that a human being does—e.g., love, feel, think, be creative—rather than just mindlessly manipulate symbols according to an algorithm. There is something barren about this way of thinking, which Rosen deftly highlighted: "Genetic engineers, who are the molecular biologists turned technologues, habitually regard their favorite organism, *E. coli*, as a simple vending machine: insert the right token, press the right button, and the desired product is automatically delivered, neatly packaged and ready for harvest."[6]

More than barren, the reductionist, formalist framework had already been detonated by Kurt Gödel in 1931 with his Incompleteness Theorem. The theorem shows that "no axiomatization can determine the whole truth and nothing but the truth concerning arithmetic." That is, Gödel demonstrated that every mathematical system must, at one or more points, rest on unproven, and unprovable axioms. The proof wasn't meant as an argument for an immobilizing skepticism. Gödel was confident that we can know certain unprovable axioms, but he was able to demonstrate that such knowledge arrives mysteriously, from outside the axiomatic system. Is that important? Roger Penrose calls it "the most important theorem in mathematical logic of all time,"[7] and philosopher A. W. Moore writes, "It took one's breath away."[8]

Is the bacterium *E. coli* completely described by the algorithmic operation of its simple genome? Or does even this basic form of life depend on something beyond the numerical or physical realm to account for the fact that it *lives*? Robert Rosen died in 1998 and his daughter, Judith Rosen, compiled a collected work, *Essays on Life*

Itself (Columbia University Press, 1999). In the dedication she wrote for the book, she asserted that the problem her father had pursued throughout his life was "What makes living things alive?," and hinted mysteriously that not all his conclusions were represented between the two covers, since "knowledge can be abused or perverted." One conclusion she said that he drew was "not everyone in science wants to 'follow the problem' where it leads or hear the truth."[9] It was a realization that Sternberg and other ID proponents would articulate many times over.

Rosen's style, too, is mysterious, as when he says, "I claim that the Gödelian noncomputability results are a symptom, arising with mathematics itself, indicating that we are trying to solve problems in too limited a universe of discourse."[10] The genetic algorithm wasn't all there is.[11] Rosen looked back to the question posed by Erwin Schrödinger more than half a century earlier, *What Is Life?* The 1944 book's "true messages, subtly understated as they are, are heterodox in the extreme," comments Sternberg. The standard, conventional "universe of discourse," the way of thinking, was far too narrow. The more capacious sort of discourse needed, it seems, would not exclude Plato's conception of transcendent ideas being involved in forming life. In Rosen's words, "A typical empiricist (not just a biologist) will say that the Schrödinger question is a throwback to Platonic Idealism and hence outside the pale of science." Thus, according to the narrow discourse, it's a view that need not be taken seriously. But Schrödinger himself meant it in earnest: Rosen says, "I think (and his own subsequent arguments unmistakably indicate) that, to the contrary, he was perfectly serious."[12] Concluded Rosen, "The consignment of the essay to the realm of archive is premature; indeed, it is again time that 'everybody read Schrödinger.'"[13]

Sternberg, studying at Binghamton University, took the injunction to heart. He found the work of cybernetician Lars Löfgren, who in a 1968 paper in the *Bulletin of Mathematical Biophysics*, asserted, "If we ask not merely how a cell can reproduce in a suitable surrounding, but how this property has evolved, then we are faced with an explanation of reproduction in a complete sense. Our results show that such a theory of evolution cannot be derived with an ordinary logical-mathematical-biological reasoning, but that it instead will have to contain new and independent axioms."[14]

What were those mysterious "new and independent axioms" that transcend "ordinary logical-mathematical-biological reasoning"? Such questions were explored by two scientists, developmental biologist Conrad Hal Waddington and French mathematician and topologist René Thom. In *The Strategy of the Genes* (1957), says Sternberg, "Waddington says, really what biologists are talking about when they're thinking of development are these very complex, multi-dimensional systems. He says the problem is biologists don't like to think that way. Mathematicians do; biologists don't." So Waddington formed a partnership with Thom, "who had the mathematical chops to actually explore this."[15]

Thom's book *Structural Stability and Morphogenesis* (1972, English translation 1975) opens with a self-evidently Platonic depiction of the universe, posing the question of why there are stable forms of any kind at all. "One of the central problems studied by mankind is the problem of the succession of forms," he writes. "Whatever is the ultimate nature of reality…, it is indisputable that our universe is not chaos."[16] So then, why order and not chaos?

Thom moves from mystery to mystery: "We perceive beings, objects, things to which we give names," he writes. "These beings or things are forms or structures endowed with a degree of stability: they take up some part of space and last for some period of time. Moreover, although a given object can exist in many different guises, we never fail to recognize it; this recognition of the same object in the infinite multiplicity of its manifestations is, in itself, a problem."

For example, I can look at a lean sloth in the wild, a corpulent sloth in the zoo, and a succession of stuffed toy animal sloths available for purchase on Amazon, based on Disney characters or other fanciful inspirations, and all are recognizable as belonging to the narrow category of sloths and sloth facsimiles, despite the latter being only whimsical likenesses of living sloths, and made out of very different materials from their living, glacially paced inspirations.

Whence this unity in diversity? It's a question that has exercised philosophers from the beginning. Thom explored the mathematics behind it, which is not simple. An overall point is that the granting of form to matter is not rigorously deterministic, as it would be if a recipe-like algorithm were at work. It is looser than that, and yet things—sloths, serpents, stars—cohere wonderfully. At the same time, "the same local situation can give birth to apparently different outcomes under the influence of unknown or unobservable factors."[17]

Those mysterious "unknown or unobservable factors" remind us of Löfgren's call for "new and independent axioms." What are they? Sternberg read Thom's book and found that it says "some shocking things." Thom is "trying to get to what is morphogenesis and what reproduction is. And he's doing it from the standpoint of differential geometry and these high-dimensional spaces that Waddington intuits, but Waddington does not have the [mathematical] tools to adumbrate." As Sternberg paraphrases, Thom says things like, "It may seem difficult to accept the idea that a sequence of stable transformations of our space-time could be directed or programmed by an ongoing center consisting of an algebraic structure outside space-time itself."

Thom "means going from a fertilized egg cell to an adult that produces gametes that then can make fertilized egg cells and so on and so forth," says Sternberg. "He's saying that you cannot explain this by a standard materialistic notion."

But there is another problem with the materialist model. Organisms are self-reproducing, unlike machines. As Sternberg says, when you start analyzing the requirements for a self-reproducing entity, it's not just that it's obviously more sophisticated than an equivalent version that cannot self-reproduce. More than this, there may be formal, even mathematical reasons just from the capacity to self-replicate that point to incalculable complexity far beyond what is in the material genome.

Eric Anderson has a chapter on this in *Evolution and Intelligent Design in a Nutshell*, where he approaches the issue from an engineering perspective. He doesn't argue for an immaterial genome but rather to the necessary sophistication of even the simplest self-reproducing cell. We must not forget the dizzying high-tech sophistication of life's capacity to self-replicate.[18]

As Robert J. Marks notes in *Non-Computable You*, computer pioneer John von Neumann developed a mathematical argument that robots could be built that could build other robots, *ad infinitum*. Could they really? A lot of proof-of-concept engineering work gets labeled as self-replicating when it's clearly far from it. Anderson shows the great gulf between "near misses" in this regard and what would be required for true self-replication.

Sternberg is familiar with von Neumann's work on the subject, as well as Rosen's engagement with that work. "When he looked at von Neumann's model in particular of a self-replicating automaton," comments Sternberg, "and of course von Neumann tried to extract formal genes from that, Rosen became clear that as it had been presented, it entails a logical paradox. The paradox is of the which-came-first-the-chicken-or-the-egg sort, attaching to self-replication: On the one hand, the components that are used to build the automaton, and that includes the description of the automaton, have to come first. On the other hand, the machine that uses those components also has to come first. And others who were sniffing around the problem, such as Eugene Wigner, also came to the conclusion that there were some serious issues with this."

From these considerations and from the observed phenomenon of self-reproduction, Sternberg came to strongly suspect that informational resources far beyond the material genome are required.

Although Thom puts the matter in modern terms, adds Sternberg, "if you read Plato and Aristotle and many of the other ancients, they were saying exactly the same thing." Sternberg refers back to the German vitalist Hans Driesch (1867–1941) who adopted from Aristotle in *On the Soul* and from Goethe the idea of entelechy, or "actualization."

"Instead of coming up with some new radical notions," Sternberg concludes, "we're back to the notion of entelechy of Aristotle, Goethe, and Driesch."

Sternberg emphasizes the importance of excavating the history of key terminology. "We often buy into terms, and we have no clue where those terms came from," he says. "They didn't fall from the sky. Often, they've been incubated for millennia. And then when you add to it a particular ideology or a particular metaphysical framework, they can become concretized and very hard to question." A term or idea that has been incubating for millennia is entelechy, that potential in life that is in a process of being actualized. He cites the theoretical biologist and logical positivist Joseph Henry Woodger, whose *Biology and Language* (1952) examines the language that animates the study of genetics, among other subjects.

It's "another book that is extremely hard to read," Sternberg admits, as the books that he recommends tend to be. "Woodger subjected genetic theories to a logical positivist analysis. He basically put everything in abstract terms, put it through a logic machine, and once defined, it didn't hold together," says Sternberg. "Most of what passes for population genetics theory, for example, is riddled with lacunae. There are holes everywhere, and many of the equations don't really work. Woodger examined genetic terms, how biologists use language, and how they often claim, 'What we've come up with is as rigorous as anything to be found in the physics.' He found that was actually not the case."

Woodger also found that different terms could turn out to mean nearly the very same thing, leading to confusion. "Woodger compared Driesch's notion of entelechy with Thomas Hunt Morgan's ideas of genetic factors, which of course Morgan said could be explained simply on the basis of mathematics," says Sternberg. "It was just a difference in terminology. Don't like the word entelechy? Use genotype. Don't like the word genotype? Use entelechy."

What all this adds up to is that a tradition has persisted from ancient to modern times which, while using different terms, has seen in the genome not a material entity alone but an immaterial one as well: abstract, mathematical, not restricted to space-time. True, the majority view among philosophers is that abstract mathematical entities don't make things happen in the material world. As will become clear, though, in Sternberg's framework, some purposive agency is driving or informing things. It's not just the "math."

While such a view will strike many as beyond the pale, it is the conventional materialist ideas of the gene, of the type critiqued by Sternberg, that has been getting knocked around by empirical findings of late. The view suffered a particularly public blow in 2012 after the ENCODE results were published. It's been called the "demise of the gene,"[19] but the aftermath has been more like a radical reconceptualization. One dramatic statement on the subject from 2024 is by Denis Noble, emeritus Oxford University biologist, writing in the journal *Nature* under the headline, "It's Time to Admit That Genes Are Not the Blueprint for Life."[20] Or see Ken Richardson writing for *Nautilus*, "It's the End of the Gene as We Know It."[21] No longer are genes understood as "discrete units distributed along a chromosome in a linear manner," says Sternberg, pointing to a 2012 article in the journal *Genome Research* by John Stamatoyannopoulos (the name is not a typo) at the University of Washington School of Medicine.[22] Instead, they have assumed a shape very much like the one sketched in this book. The gene "represents a higher-order framework around which individual transcripts, or RNAs, coalesce, creating a poly-functional entity that assumes different forms under different cellular states, guided by differential utilization of regulatory DNA."

A "higher-order framework" is very far from particles arranged on a string. "It has to be more than an invariant particle, almost by definition," says Sternberg.

What empirical evidence do we have for such a view? Sternberg outlines four principal clues.

8. Clue One:
The Flux Computer

T HE 1985 COMEDY ADVENTURE *BACK TO THE FUTURE* INVOKED A mysterious "flux capacitor" as part of the time machine used by the story's hero. The device is, of course, fictitious. What isn't fictitious is what we might term a "flux computer," a technology so advanced that when described it sounds like science fiction, but is in fact all around us, and in us.

This brings us to the first of Sternberg's four clues pointing to the immaterial nature of the genome.

We begin to see it by observing the curious way that coded information in DNA is characterized by interleaving. It overlaps with itself. In plants and animals, particularly in mammals, says Sternberg, "a segment of DNA consists of interleaved, interspersed, multilevel, and overlapping data files." The gene for this and the gene for that are "interleaved, interspersed, and chock-full of other DNA elements." The overlapping makes possible "logical circuits" as in a computer processor. The organization is seemingly madcap, with elements that code for proteins mixed together with those that don't, and with other DNA elements. Sternberg lists promoters, silencers, super-enhancers, micro-RNA genes, virus-like elements, transcriptional terminator sequences, imprinting elements, and more. Don't worry about what all those are. It doesn't matter for the argument beyond their helping us recognize the stunning sophistication of the system.

The comparison with a computer sounds like we are headed for a more familiar design argument, one that would cite Bill Gates in his book *The Road Ahead*: "DNA is like a computer program but far, far more advanced than any software ever created."[1] The key here, though, is not what makes DNA like a computer but what makes it unlike a computer—what makes it beyond a computer. "The difference that makes the difference is that this kind of circuitry is not fixed like on a standard processor we would use," explains Sternberg. "It's dynamic. It's changing constantly."

The flux computer.

DNA produces RNA transcripts, but genes can produce not one but hundreds or thousands of different transcripts. The details are frankly bewildering, as it can be to listen to Sternberg discuss them. But the system—with "what I'm calling data files," says Sternberg, "formed into folders, as I would call it," and further into "superfolders" arranged in "banding patterns"—is not random. It is hierarchical. And yet the overall impression for us is a chaotic haze: "When you try to pin down what exactly a gene is in this larger context, it becomes difficult because where one starts and where one stops is very hard to define. It becomes blurry. It becomes fuzzy."

You could think of it this way: It is like a library but the craziest library, filled with the craziest books that you can possibly imagine, and wildly vast in scope. But that's not all. The library is delivering books to individual users and rewriting books or sections of them to tailor fit the books to different library patrons (for example, reducing the suspense level for someone who prefers a lighter read, changing the ending for someone who prefers a happy over a hopeless one, or lowering the reading level of an adult book checked out by a kid). All this is happening at lightning speed. For the cell's needs, something is reading and revising the information in this library on the fly. But how?

Whatever the answer, the life of every living creature depends on its being able to do so. Last night I was outside with a flashlight after a rain, looking for food to feed to my snail. His favorite is arugula, which he gobbles with remarkable speed, but he will settle for lichen. The flashlight fell on the soil around a tree. I noticed small dark forms leaping into tiny holes in the ground. They were earthworms, disturbed by the light. It was startling to see how fast they moved, and with what accuracy in their blind dive for cover. (Earthworms are blind but can detect the presence of light on their skin.) How do they manage such a feat? It is contingent, at every moment, on that mysterious reader of the physical genome, reading and sorting at lightning speed. There's a method to the madness, and your life depends on it every bit as much as an earthworm's does.

That lightning-swift reader in the madcap library suggests not just purpose—say, a designer's purpose acting in the past, whether at the origin of life billions of years ago or in the numerous "big bang"-type pulses of biological novelty such as we find in the Cambrian explosion. It suggests, says Sternberg, purposeful agency acting right now.

That is the first clue for the immaterial genome.

9. CLUE TWO:
CHOOSE YOUR OWN ADVENTURE

THE HUMBLE TIGER PRAWN, *PENAEUS MONODON*, IS A SUPERMARKET staple with an oddly intellectual expression on its face. The crustacean relates to Sternberg's second clue about the immaterial genome, this one involving RNA transcripts.

According to the simplistic formula you may recall from high school, DNA codes for RNA (or messenger RNA, mRNA), which codes for proteins, which codes for you, or for tiger prawns, as the case may be. The first of these steps, from DNA to mRNA, is called transcription, pointing up the fact that the process produces a transcript of sorts.

Genes can encode many transcripts, but the tiger prawn holds a record as far as Sternberg knows. He says a single location (or "locus") in the tiger prawn genome, the DSCAM locus—short for "Down syndrome cell adhesion molecule," since in humans it is associated with Down syndrome—is "266,000 base pairs long, but it has the potential to encode no fewer than 21 million different protein isoforms" (isoforms, meaning similar but distinct proteins). Before learning this about the tiger prawn, Sternberg had understood the fruit fly *Drosophila* to be the champion, with a DSCAM locus encoding 38,000 or more protein isoforms.

Operations on RNA going on in the cell—RNA editing of various kinds—result in a bewildering array of products. "The point I'm making," says Sternberg, "is that a gene in this context provides a substrate for many types of information that are layered on, so to speak, by the cell because the RNA is, yes indeed, a message of a sort. But it's cellular processes that are coming in and changing the code script. And because of this, many RNAs do not mirror any particular DNA sequence. And there's some extreme cases of that." It's like an orchestra, of immense size and complexity, with instruments of many, many kinds, playing a hugely complex musical score. The idea is similar to Denis Noble's in his book *The Music of Life: Biology Beyond Genes*.[1] But who is at the podium looking at the music on the stand before him and conducting? And where is the score? The orchestra plays near-perfectly. Yet the score keeps changing. Also, individual orchestra members often take breaks and are replaced by other performers. They switch from musical performance to other activities—perhaps painting portraits or cooking fine meals for each other and for the audience. Who or what is coordinating all of this activity?

Sternberg notes that the code sequence of the RNA in many cases isn't the point; instead, its shape is what's important: "The RNA-level codes that are formed are often topological or rather geometrical in nature, meaning that they form complex shapes. These are recognized by various proteins, and they can have metabolic functions too. And many of these RNA-level codes are sequence independent. You can change the sequence dramatically, and yet they will perform the same function."

The orchestra analogy breaks down—analogies always do; otherwise they wouldn't be analogies—but perhaps that one gives a rough sense of the processes going on. Or, returning to the image of a library, another comparison, this one suggested by Emily Sandico, is to the old "choose your own adventure" books some of us read as kids. These books gave you a choice at certain junctures. Depending on the reader's preference, the story's protagonist could make one choice or another, take one action or another. The reader would then turn to an indicated page, say 32 or 54, where the adventure would carry on based on that choice. A book like that includes many possible outcomes. But who is doing the choosing? Obviously, it's the reader. Now imagine a whole enormous library of choose-your-own-adventure books, all of them being read by some outside agent. The reader is not itself the library. And without the reader, the library would be inert. It would tell no stories at all. RNA editing, argues Sternberg, entails a reader. But who or what would that be?

It breaks one's imagination to think all this is happening seamlessly without direction. This is the second clue pointing to the immaterial genome.

10. Clue Three:
Garbage Islands No More

THE THIRD CLUE RELATES TO SO-CALLED "JUNK DNA," A MYTH duly exposed as long ago as 2011 by, among others, Richard Sternberg's biologist colleague Jonathan Wells. (See Wells's *The Myth of Junk DNA*.) Evolutionary biologists had confidently asserted that the vast majority of DNA consists of the sort of debris you would expect to accumulate from an unguided, purposeless process. It's like one of those garbage islands floating in our oceans, with refuse from years of human mismanagement of unwanted plastic and other materials. But as noted in an earlier chapter, the vast majority of the "junk" is active in some way, being transcribed to RNA. If it had no function, it's highly unlikely it would be transcribed. Why expend that energy for nothing? And indeed, many functions for the supposedly function-less junk have been discovered.[1]

But this clue is about more than functionality as such. After all, as we saw earlier with Richard Dawkins, a smooth Darwinist can change his tune on that subject in a trice, from "Based on evolutionary principles, we expected it to be mostly junk," to "Actually, based on the very same evolutionary principles, we expected it to be mostly functional."

Rather than function alone, this clue is about the patterns in which repetitive DNA and other non-junky "junk" appear in the chromosomes. Sternberg reads these patterns like a Hercule Poirot. The distribution is not random but instead seems to be telling us something. More on this in just a moment, after we lay some groundwork.

"In the case of humans," says Sternberg, "it's roughly 98 percent of our DNA that does not code for protein. Of that 98 percent, some estimates would be that a maximum of 10 percent can have some function,[2] based on estimates of sequence conservation and based on other properties." The vocabulary relating to DNA elements that do not (or seem not to) code for protein is arcane. In *The Myth of Junk DNA*, Jonathan Wells offers some crisp, clear definitions of relevant terms:

Conserved sequences: DNA or RNA sequences that are similar in different organisms. According to evolutionary theory, if two lineages diverge from a common ancestor that possesses DNA sequences that are nonfunctional, those sequences will accumulate mutations that render them different ("divergent") in the two descendant lineages. But if the original sequences are functional then natural selection will tend to weed out mutations, and the corresponding sequences in the two descendant lineages will remain similar ("conserved").

Repetitive DNA: A DNA sequence that is repeated in the genome—in some cases, thousands of times. About half of the human genome consists of repetitive DNA, and about two-thirds of those repetitive sequences are LINEs or SINEs.

LINE: Long Interspersed Nuclear Element, a retrotransposon and one type of repetitive DNA. LINEs tend to be more than 5,000 nucleotides long and include a DNA sequence encoding an enzyme that enables them to reinsert themselves into DNA. Mammalian genomes contain tens of thousands of LINEs that fall into several groups; the most common of these is called L1.

SINE: Short Interspersed Nuclear Element, a retrotransposon and one type of repetitive DNA. SINEs tend to be less than 500 nucleotides long and depend on other mobile genetic elements for their retrotransposition. The most common SINEs in primates are *Alu*s; rodent genomes contain different SINEs called B1 and B2.

Transposon: A mobile genetic element (known colloquially as a "jumping gene") that can move from one place in the genome to another, in what amounts to a "cut and paste" process.

Retrotransposon: A mobile genetic element (transposon) that uses RNA as an intermediate in what amounts to a "copy and paste" process. The DNA element is first transcribed into RNA, then an enzyme called reverse transcriptase copies the RNA sequence back into DNA that is inserted into a different place in the genome.

Pseudogene: A non-protein-coding segment of DNA with a nucleotide sequence that resembles a DNA segment that codes for protein elsewhere in the same organism or in other organisms.

ERV: Endogenous RetroVirus, a genomic sequence that resembles (and might be derived from) the sequence of an RNA virus that has been reverse transcribed into DNA.

Reverse transcription: A process in which the nucleotide sequence in a strand of RNA is copied into DNA, catalyzed by an enzyme called "reverse transcriptase."[3]

Sternberg speaks of chromosome "banding," a technique of staining chromosomes. "When you band chromosomes using LINEs and SINEs, you get these very distinct barcoding-like patterns." These patterns, he says, tell us something important. Referring back to his first clue for the immaterial genome, he says, "When we look at how so-called non-conserved 'junk' sequences are dispersed, what we find is that they're part and parcel, they're warp and weft, of this interleaved, interspersed, multilevel, non-random organization."

Sternberg alludes to Francis Collins and his book *The Language of God*, published in 2006. He paraphrases Collins: "Well, there's just so much junk DNA out there and there's no logic in the madness." Collins was mistaken. Ironically, his name was among the co-authors of a 2004 paper in the journal *Nature* that guided Sternberg's thinking:

"Genome Sequence of the Brown Norway Rat Yields Insights into Mammalian Evolution."[74]

Sternberg compares this rat to a mouse. "Now the interesting thing is that the SINE content of the brown Norwegian rat is different from that of the mouse. But if you look at the distribution patterns, they're remarkably similar. It looks almost as if it were genomic plagiarism, as if you just drew one graph and then decided you were going to make a little bit of variation but give the same graph again."

He compares SINEs in rodents and humans. "These sequences"— B1 and B2 in the former, *Alu*s in the latter—"have originated independently of one another," he explains, summarizing the mainstream view among evolutionary biologists. "They haven't been passed on by common descent. Yet, if you look at their distributions along chromosomes, what you find is that their patterns are remarkably similar."

Sternberg and University of Chicago geneticist James Shapiro found in their collaboration that there is function in repetitive DNA elements—significantly, when and how genes are expressed during the embryo's development. How did they arrive at such a conclusion? "You have these non-random patterns," says Sternberg. "They look very similar. The elements themselves are not similar. Their distribution patterns are similar."

Poirot, Agatha Christie's Belgian detective, would appreciate what Sternberg is doing here. I can hear Poirot saying that before trying to read a language, you must first recognize that it is a language. If the "junk" elements themselves are different but are distributed along the chromosomes similarly, that looks intentional, like a language. The barcoding on the packages in your grocery cart is a language, but is read by the scanner when you check out at the register, not by a human being. To our eyes, the barcoding looks meaningless, but it's not.

In the biological context, because the sequences are not identical— i.e., it is not always the same type of SINE—the striking similarity of the two barcodes can't be explained by common descent. That is a conclusion of the 2004 *Nature* paper mentioned just above, which says that they must have been inserted "independently." The similarity can be explained, as with any language, by purposeful agency. If the purpose is to shape development, then an agency over and above the physical aspects of the genome—over and above the sequences of DNA—is doing that.

This is the third clue for the immaterial genome.

11. Clue Four:
A See-in-the-Dark Rat

"Imagine a rat," Sternberg is saying. "It lives in the recesses of the New York City subway system, far, far, far below ground, where photons are few and far between. And yet the rat is able to see to some extent." How is that possible? "That's because its rod photoreceptor cells are adapted to do so."

This is the fourth clue. The organization of what was thought to be junk DNA is not only non-random. It is functional in surprising ways relating to the architecture of the cell's nucleus with its chromatin and other elements. The arcane details include chromosome territories and topologically associating domains. To take the latter instance, topologically associating domains are basically sections of different chromosomes that reside near one another when all the chromosomes are bunched up in the nucleus of a cell. This allows them to interact biochemically. Think of it like this: Chromosomes are all coiled in a three-dimensional shape in the cell; and what's bunched against what is not inconsequential. For example, section 3,942 on chromosome 8 might need to be placed near section 40,912 on chromosome 15 because there's an enhancer in that locus on chromosome 8 that regulates the production of a gene in the corresponding locus on chromosome 15. Chromosomes, in other words, interact with one another in highly complex and specified ways that are hard to predict, ways far more complex than molecular biologists had realized.

A few more terms, defined by Jonathan Wells:[1]

Chromatin: The combination of DNA, proteins, and RNA that makes up a chromosome. It includes histones, special protein spools around which the DNA molecule is wound.

Satellite DNA: A fraction of DNA consisting of millions of short, repeated nucleotide sequences that produce "satellite" bands when DNA is centrifuged to separate it into fractions with different densities.

Tandem repeat: A form of repetitive DNA in which (usually short) sequences of nucleotides are repeated adjacent to each other. Satellite DNA consists of tandem repeats.

In the case of nocturnal mammals like our New York rat, the chromatin functions in a very special way. Says Sternberg, "Look at where the sequences are organized in the nucleus" (the DNA sequences that do not code for proteins). "Look at the positioning of SINE elements, and the positioning of LINE elements, and the positioning of what are called satellite DNAs. These are tandem repeats—it would be like 100 base pairs, repeated over and over again. They have distinct nuclear localizations, different from most other cell types in the rod photoreceptor cells of these mammals. The chromosomes are positioned in such a way that the chromatin is acting as an optical device."

An optical device, fashioned from "junk DNA," without which the rat's underground lifestyle would be impossible. Chromatin (which is basically chromosomes when they are all coiled and wrapped up in proteins and other molecules) in these cells helps rats see in the dark. The details of how this happens are beyond our scope—how chromosomal regions with different optico-physical qualities are sorted to optimize in 3-D the scavenging of photons, in effect building a super-light-sensitive receptor. But the existence of such phenomena casts non-coding DNA sequences in another light altogether. Chromosomes do a lot more than we thought. Coding for proteins is just a part of it.

In light of such findings, we have to revise our understanding of the physical genome, and not just in the details, as fascinating as those are. We need to take better account of the gene as a "poly-functional entity" and how this further highlights the inescapable reality of the physical genome's immaterial counterpart.

I asked Sternberg if perhaps instead these intriguing findings regarding the chromatin of nocturnal mammals are just evidence for an additional layer of software sophistication in the genome. His answer was, in effect, listen to what you yourself have said: "That thread of computation is what should be focused on, as opposed to the DNA substratum it's coming from. If you think it's just another layer of software, what is software? It's a completely immaterial entity, reflected in zeros and ones. Everything depends on numbers. The hardware is important but secondary. As soon as you talk about software, you are talking about something immaterial, unless you want to suggest that numbers are material objects."

He offers me off the top of his head several papers indicating that information as such, as in software, must be immaterial, including two bluntly titled ones: "Information Is Not Physical," by Robert Alicki,[2] and "Information Is Non-Physical," by Richard Liangchen Wang.[3] And a third by Pentti Kanerva on "Hyperdimensional Computing," which states, "The materials that a circuit is made of are incidental— we could say, immaterial."[4]

Sternberg explains more about Kanerva, a neuroscientist: "He goes on to say that the lesson to be learned from the study of computers is that any logical design can be separated from its physical actualization. For any computation can be executed using either media generated from metal-oxides and silica, or with the components of a cell. In other words, the supposition that *algorithm* = *gene* brings with it the implicit notion that such an object is matter-independent as a paradigmatic cause. As related to a 'gene' object, the use of *algorithm, blueprint, code, coding* or *encoding, data, information, instructions, procedure, program, recursion,* and *subroutine* as descriptive terms imply, then, a form of independence from a substrate."

And for dessert, Sternberg quotes Roger Penrose: "Like so many other mathematical ideas, especially the more profoundly beautiful and fundamental ones, the idea of computability seems to have a kind of *Platonic reality* of its own" (emphasis added).[5]

This—the immaterial nature of the information guiding heredity, guiding biological development, guiding life—is Sternberg's fourth clue pointing to the immaterial genome.

12. An Evening's Entertainment

RICHARD STERNBERG REFERS TO HIS RESEARCH WITH GENETICIST James Shapiro and their reconceptualization of what DNA really is. Arising from his work with Shapiro was the realization that, "instead of seeing DNA as this master molecule calling the shots, another way of seeing it is as... a very interesting, information-rich data storage medium," but one that is being operated by something else—something using the DNA.

This marks a revolutionary break from Francis Crick's central dogma of 1958. Crick wrote that "once 'information' has passed into protein *it cannot get out again*. In more detail, the transfer of information from nucleic acid to nucleic acid, or from nucleic acid to protein may be possible, but transfer from protein to protein, or from protein to nucleic acid is impossible. Information here means the *precise* determination of sequence, either of bases in the nucleic acid or of amino acid residues in the protein."[1] Sternberg summarizes this in simpler terms: "DNA makes RNA makes protein makes us." In the same 1958 paper, Crick stated the famed sequence hypothesis: "In its simplest form it assumes that the specificity of a piece of nucleic acid is expressed solely by the sequence of its bases, and that this sequence is a (simple) code for the amino acid sequence of a particular protein."[2]

According to the central dogma outlined by Crick, DNA as the "master molecule" is fixed—that is, its sequence of nucleotides never changes, absent some accidental mutation. But the new understanding sees in it something far more dynamic. "It is not just read but can be written," says Sternberg. This explains why DNA is not exactly the same in all the cells in your body. In your brain, for instance, there exists a "mosaic in terms of DNA sequences," not wildly different but with significant variations. The same is true of fruit flies and other organisms. He says, "The lecture, later a paper in *Science*, that put cell-induced DNA-level changes front and center was Barbara McClintock's 1983 Nobel lecture, 'The Significance of Responses of the Genome to Challenge.'[3] In 1992, James Shapiro introduced the concept of 'natural genetic engineering.'[4] He continues to add new evidence and insights on the topic. A paper co-authored by Laura Landweber, titled 'What is a genome?,' discusses some organisms that literally sculpt their 'genes' in each generation." Landweber and her co-author, Aaron David Goldman, conclude, "These examples suggest a more expansive definition of the genome as an informational entity, often but not always manifest as DNA, encoding a broad set of functional possibilities that, together with other sources of information, produce and maintain the organism."[5]

This is not how we learned about it in high school, or likely college. DNA "can be changed in three dimensions and in the fourth—that is, in space *and* time—but in other ways too," says Sternberg. What other ways? He mentions "layers of structure, n-dimensional structures, Hilbert space, events of space and time that are effects of a higher-order, possibly infinitely dimensional structure." (This is how he talks in casual conversations.)

Sternberg uses the word "computational." DNA with its repetitive and other "junk" elements is an active tool, but in the hands of what or whom? The question, bracingly Platonic, is unwelcome to many people. "What bothers them," says Sternberg, "and it's been my experience that it does bother them, is that you begin to attribute an informing principle to something other than DNA." What is the nature of this informing principle—that is, other than DNA and other physical, epigenetic sources of information in the cell? "There has to be agency," he says. "There has to be something that's making the decision as to when and where it's going to be reading and using DNA in certain ways."

He gives an illustration from the world of home entertainment. You're planning to watch a movie on your laptop and, rather than using a streaming service, you have a hard drive where it's stored. But, says Sternberg, "imagine you've got many, many different movies superimposed on the same drive." It is a simple way of stating the outcome of his four clues. You sit down and deliberate over the options. Which movie will you watch? It's not the hard drive that decides. It's you as an intelligent agent outside the storage medium: "You say, 'I'm going to watch this movie at this time. Or maybe not. Maybe I'll switch to something else.' And so when you start looking at DNA that way, things begin to change."

He calls the more traditional view of Francis Crick, or of biophysicist Max Delbrück, "DNA as the unmoved mover." The "unmoved mover" is the Aristotelian way of referring to God. The Crick/Delbrück notion turns DNA into the god, a reversal of Sternberg's immaterial genome.

13. More than Machines

THE ENGLISH HISTORIAN AND NOVELIST A. N. WILSON, WHOM
I admire, says something very mistaken in his book *God's Funeral:
The Decline of Faith in Western Civilization* (1999). In the preface he
characterizes DNA as a stumbling block for religious believers. "The
Darwinian who points to the mid-twentieth century discovery of
DNA as a confirmation, beyond reasonable doubt, that the theory
of natural selection [i.e., Darwin's theory of unguided evolution] was
correct, can do nothing to alter the beliefs of the Creationists."[1] Here
I assume he is using "Creationist" in that idiosyncratically broad sense
typical of Darwinists intent on discrediting any and all opponents
with whatever unfashionable label they can find, to encompass everyone who rejects the idea of purely mindless Darwinian evolution. This
sense of the term is broad enough to encompass even those who believe
in both a billions-year-old universe and a history of life marked by
dramatic evolutionary changes, but who conclude from the evidence
that those changes were guided and made possible by a source of
intelligent agency.

One irony is that Wilson would come out as a Darwin skeptic
himself in a later book, *Charles Darwin: Victorian Mythmaker* (2017).
Another, more important irony is that DNA, far from pointing away
from God and toward materialist theories of biological origins, does
just the opposite. As we've seen so far in Richard Sternberg's story,
the deeper that science delves into the genome, the more apparent it
becomes that intelligent agency over and above the material dimension of the genome is involved. In Sternberg's argument, I find that

most clearly indicated when he turns again to a dry-sounding subject: "computational issues" that arise when attempting to understand the genome as something purely physical.

Darwinists and design theorists alike are wont to imagine structures in the cell as "machines": for example, the ribosome as a machine that forms proteins from RNA transcripts, and the spliceosome as a machine that splices RNA, distinguishing exons (which code for protein) from introns (which do not), before the RNA is processed by the ribosome. Sternberg directs our attention to the spliceosome. Another term defined by Jonathan Wells:

> **Splicing**: The process in which the exons in an RNA transcript are put back together and the introns are cut out. In alternative splicing, some exons may be omitted while others may be duplicated.[2]

A problem with describing such devices as "machines" is that the term is only a metaphor, and one that breaks down in important ways that are frequently overlooked. Recall the tiger prawn with its DSCAM locus capable of generating 21 million protein isoforms. The "machine" in this case, formed from RNAs and proteins, must select which of those similar proteins it will generate. The spliceosome can be thought of as a particularly sophisticated kind of machine, a factory, but it's qualitatively different from the factory that built, let's say, your car. An auto factory has many moving parts, but the structure of any given car factory is relatively fixed. It would have to be to function efficiently. Not so the spliceosome. "What you find is that the proteins and RNAs [that constitute it] are in a state of flux," says Sternberg. "There is no fixed machinery."

"Imagine you have a factory floor," he continues. "You open the door to the factory and you see these robots and they're involved in producing whatever, widgets. In this case, the widgets would be RNAs. You open the door, you look in, and you see a panoply of machinery going at it. You close the door, open the door some seconds or minutes later, and what you find is that all of the first group of robots is gone. Now a whole other set of machines are operating on RNAs, and they're conducting their business. And you close the door and walk away, and come back, open it again, and what you find is now the machinery has changed again."

This is the surreal reality of cellular "machinery." It is a dance, with "choreography," as Sternberg puts it, as "the factory is constantly reorganizing itself as it performs diverse operations." Indeed, the dancers themselves are metamorphosed in the process. But who or what is the choreographer that not only selects which of 21 million proteins to make but also directs the process by which they are made and employed? The metaphors all fail, but they're all we have. Another: "The components are coming in and they're coming out like actors on the stage." But not quite like that, because the actors have a set script. They are not choosing from among 21 million variants on the script. For operation of the spliceosome, there is no "static instruction set." And actors have a director. What is doing the "directing" as the tiger prawn gestates?

Isn't the instruction coded into the DNA? No, explains Sternberg: "You run into problems there because we are not aware, at least to my knowledge as of this day, of any 'direct algorithm' that is saying the protein can adopt all these various conformations"—that is, the three-dimensional shapes of the folded proteins. DNA is not an algorithm. It's not a computer program. "Now, this is remarkable because it means that you've got a series of decision-making procedures that are mediated by a meta-dynamic network," says Sternberg. And this is the key: "If you look at any one of these networks and you look at the number of components involved, and you look at the number of states that the different components can adopt, you suddenly find that the states, the number of potential configurations, are hyper-astronomical." By "hyper-astronomical," he doesn't just mean a really big number. He means a number so enormous as to pose an overwhelming challenge to any strictly materialist account of the genome's operation.

The math is relatively easy. It can be done "on the back of an envelope," says Sternberg. Here it is: "Imagine that you have, let's say, 200 different protein components, and each of those components can adopt, say, at any given time, any one of ten different states." Ten to the power of 1 is 10. Ten to the power of 2 is 10 times 10, or 100. Ten to the power of 3 is 10 times 10 times 10, or 1,000, a 1 followed by 3 zeroes. "In this case," says Sternberg, the likelihood that all 200 protein components of the network would be in a particular appropriate state given cellular or developmental conditions is "1 chance in 10 raised to the 200th, which is a very large number." That is, it would be a 1 followed by 200 zeroes.

Thus, whatever is doing the information processing must be a tremendously powerful information processor: "If you had, for example, logic circuits and each one could only be either on or off, a zero, or a one," says Sternberg. "And if you had 10 of those circuits, then how many states would you have? You would have two states raised to the 10th power. If you had 100 of those logic gates, then it would be those two raised to the 100th power." Sternberg cites the work of such men as cyberneticist W. Ross Ashby and developmental biologist Conrad Waddington. They sought to understand the complexity of a brain—let's say a tiger prawn's brain or an artificially intelligent brain. If you were to say, "Well, there's a program running all of this," then, says Sternberg, "the informational capacity of that program has to be equal to, or greater than, the number of states you have to process."

Sternberg refers to computational modules in "complex dynamical systems." "If you say, 'I've got 10 states here and I've got 2,000 different modules,' you come up with a very, very large number." That is, 10 to the 2,000th power, or a 1 followed by 2,000 zeroes. That is the number of possibilities. And as Sternberg further notes, "Whatever is controlling it has to be, in its informational capacity, equal to or greater than the number of possibilities. Ashby referred to this as the law of requisite variety."

But are such hypothetical figures relevant to, say, the activities of the spliceosome? "If I start looking at it in terms of numbers," says Sternberg, "it becomes less and less likely to be a standard run-of-the-mill mechanical process," the sort necessarily entailed in the model of the genome as a purely physical entity. The numbers are not just large but "transcomputational," he says, beyond the capacity of any purely physical system to compute in our material universe. The German mathematician and computer scientist Hans-Joachim Bremermann (1926–1996) defined the limit of this. He "imagined a computer the size of the Earth, of which every atom is devoted to processing bits, running for however long the Earth has been in existence. So, billions of years. And his question was, how many bits could it process? He calculated that it would be a maximum of about 10 to the 93 bits." That is, 1 followed by 93 zeroes.

That's a problem, because in the examples Sternberg gives of normal activities going on in the nucleus of a cell, the information processing is already far beyond the number calculated by Bremermann. It would seem that the nucleus is not of the age or size as a computing entity to manage the computations expected of it. To say "not even close" is an absurd underrepresentation of the problem.

One of Sternberg's professors at Binghamton University, computer scientist George Klir (1932–2016), "devoted some ink to Bremermann's limit and the transcomputationality problem," says Sternberg. "If you can take standard genetic regulatory networks, and if you look at the number of bits involved, it's not hard to get past 10 to the 93 bits very, very quickly." As for Sternberg's example mentioned above, 10 to 2,000th possibilities, how many bits are involved? He explains, "The length of any given bit string (or state) is not the problem. The problem is searching for the very small subset of those bit strings that reside in a space of 10 to 2,000th possibilities. The problem is finding a restricted subset of those bit strings, maybe only one bit string, that gives a desired outcome." What you are doing right now, your brain processing light to represent images decipherable as text to be read, is an example of a system at work that breaks Bremermann's limit.

Mechanical engineer Seth Lloyd at MIT has likewise sought, in a paper in the journal *Nature*, to find the "ultimate physical limits to computation"—that is, of our entire universe, assuming access only to purely physical processes.[3] Lloyd, says Sternberg, "estimates an upper limit of around 10 to the 120th, in terms of bits." A higher number calculated by intelligent design theorist William Dembski, what Dembski labeled the "universal probability bound," is 10 to the 150th. "That also is a number that," says Sternberg, "when you're dealing with these regulatory networks, is not hard to get to. It becomes even more so if you think of the processes occurring as being analog versus digital, where you're dealing with a continuum of states. And then it just shoots off very quickly into some countable infinity."

And this is not just a matter of something that happened at the origin of life, or at the Cambrian explosion. It must happen during development, as each embryo grows. And not only that. The "informational capacity that can regulate a system like the spliceosome, like the ribosome," must be "somehow ongoing," says Sternberg, and "whatever's regulating it has to be bigger than, or at least equal to, all the problems that would potentially have to be solved."

As is his habit, Sternberg harks back to great scientists of the past. "Certainly Wilhelm Johannsen and William Bateson, and others, would not have doubted it. Whatever agency it is that's controlling these processes in space and time, going from one cell to two cells, to four cells, to eight cells, on to the trillions of cells to make up a human being, the differential reading of these DNA sequences has to have a capacity that I think is unfathomable." The numbers cited by Bremermann, Dembski, and Lloyd don't even come close. The number of potential states for a simple organism such as yeast is thought to be "10 to the 79 billion." That's the number 1 followed by 79 billon zeroes. Thus, "You would have to take those potentials and you would have to map them onto the yeast chromosomal DNA sequence." With higher organisms such as mammals, the problem obviously becomes enormously more complicated.

This would seem to suggest that the building of a developing animal must be guided by more than just matter in motion. Something in the process must be immaterial. If we call that something the genome, then we have in view an *immaterial* genome.

14. Finitizing the Infinite

The way forward is the way back.
—David P. Goldman[1]

Thinking about an immaterial genome, says Sternberg, "scares the hell out of a lot of people." It does, because it forces us to think in terms of philosophy, or theology, not of a dry kind but of an eerie one. The computations in life can't be purely physical. Therefore that leaves the spiritual, perhaps even the supernatural. What else shall we call it when it transcends the limits of our natural world? I'm open to suggestions but no alternative occurs to me. I also don't see an alternative to calling it eerie.

Sternberg refers to a "countable infinity," an idea from mathematician Georg Cantor and other exponents of set theory. Somehow, the infinite, or some order of infinity, is operating in each finite life, and each finite cell within it.

I was startled when I came across a relevant essay on this topic by a friend of mine, the writer and economist David P. Goldman. Goldman in recent years has emerged as a provocative spokesman for the theology of Rabbi Joseph B. Soloveitchik (1903–1993), the most prominent American Orthodox Jewish thinker of the twentieth century. This is not a book about Goldman or Soloveitchik, about the Bible or the Jewish mystical tradition known as the *kabbalah*,[2] an interest of Sternberg's. I just want to note the overlap.

Goldman has written about Cantor and the orders of infinites. He has thought about how "the antinomies set forth by Parmenides in the eponymous Plato dialogue are instantly recognizable in the paradoxes of modern set theory."[3] And he has highlighted Soloveitchik's writing and teaching on *tsimtsum*, the kabbalistic idea that the infinite God, in creating the world, contracted within himself in order to make space for the finite. Quoting Soloveitchik: "It is Judaism that has given the world the secret of *tsimtsum*, of 'contraction,' contraction of the infinite within the finite, the transcendent within the concrete, the supernal within the empirical, and the divine within the realm of reality. When the Holy One, blessed be He, descended on Mt. Sinai, He set an eternally binding precedent that it is God who descends to man, not man who ascends to God."[4]

According to a contemporary interpreter, Meir Triebitz, *tsimtsum* answers the question: "If I begin with God and God is all there is, then by definition He encompasses and defines all that there is. If so, how can God possibly create something which is 'outside' of Himself?" He also notes that this is "not a process of literal Divine contraction, but rather a contraction of God's thought or will."[5] Jewish tradition finds this idea alluded to in the first word of Genesis, "*Bereishit*," usually translated as "In the beginning," though the expression is more multivalent than that. In the same essay, Goldman observes:

> The finitization of the infinite, the "secret" that Judaism imparted to the world through *tsimtsum*, is the defining characteristic of what for lack of a better word we call the "modern." In the middle of the 15th century, the West began to see the world differently thanks to perspective in painting, and to hear the world differently thanks to tonal counterpoint in music. By the middle of the 17th century, we understood the universe in an entirely new way, through the laws of planetary motion and infinitesimal calculus. All of this presumes a Creator God who makes this engagement possible by contracting his infinitude.[6]

The "finitization of the infinite" is exactly the problem that Sternberg wrestles with, and so it seems did Darwin. In an essay on the immaterial genome that Sternberg shared with me, he notes that Darwin "surmised that his 'gemmules' (which later became termed 'pangenes' in 1892 and then 'genes' in 1909) had to be *nearly infinite in number to effect morphogenesis,*" that is, to bring about the generation of organismal forms. Darwin wrote in a manuscript ("Hypothesis of Pangenesis"), unpublished during his lifetime (though the contents made it into his 1868 work *The Variation of Animals and Plants Under Domestication*), that "in every animal and plant there must be innumerable latent, self-propagating gemmules, ready under fitting circumstances to be developed." And "each organic being may be looked at as a little universe, formed of a host of different self-propagating organisms, almost as numerous as the stars in heaven, and as minute as they are immense."[7]

It is a lovely passage. Sternberg comments that he himself uses "several theorems to show that Darwin's appraisal on this point is correct—that such a process-shape is infinitely dimensional." Sternberg asks, "Can, then, some tens of thousands of 'genetic instructions' control what is at its base an ever-expansive, non-finite orchestration? Otherwise stated, can that which is infinite be circumscribed by a finite string of symbols?"

In email correspondence, Sternberg answers his own question by returning to the metaphor of the library and its necessary reader (discussed in Chapters 8 and 9):

> DNA can be regarded as a database or "library" (following D. L. Nanney[8]) that can be used in a read/write manner. As such, then, it is necessary but not sufficient for anything that we call (sloppily) DNA-level "encodement." For coding/decoding/encoding there has to be some ontologically prior system of interpretation, which can select the data to be "read," compiled, "executed," and so forth. (E.g., if the database/library is in a language that is not interpretable, then it is the same as having no database/library at all.) So, to my way of thinking, one can properly say that a "protein-coding region" is such, but only in the sense that another linear string (RNA) can be retrieved from it, which in turn can be translated into a protein (or part thereof). To posit, though, that such a coding region is an instruction or procedure for how that protein will fold in 2D/3D/4D, and for how it will interact with other molecules, is incorrect for a host of reasons. Hence… "that which is infinite"

has NOT been "circumscribed by a finite string of symbols." Rather the infinite (this high-dimensional structure) only makes crucial use of the "finite string of symbols."

Thus are we thrown back on Goldman's subtly more modest term "contracted," borrowed from Soloveitchik. And even on this ground I believe the matter can only be understood as a paradox, much like the kabbalistic *tsimtsum*. Somehow, the infinite directs the finite by being contracted into it.

If I've understood Sternberg correctly, the new science of the immaterial genome can be compared to puzzles from scripture. Not that it's a religious idea, but men and women have been wrestling with questions like Sternberg's for millennia. It's not surprising that, as with Plato, this new science echoes ancient ideas. Christians may be reminded of their conception of God becoming a man on Earth. I think of the lowly thorn bush from within which God himself addresses Moses. It burns and yet is not consumed. God is in the "heart" of the flame, contracted, as Soloveitchik puts it, to a "dimensionless point."[9] Similarly, the Talmud infers that the Ark within the Holy of Holies, containing the tablets summarizing the covenant with the infinite, occupied no physical dimensions.[10] The parallel to organic life is this: In Sternberg's thinking, it is the presence of a transcendent infinity, necessitated by the issue of transcomputationality, that seems to be what calls the immaterial genome into being.

It would be interesting to put Goldman in the same room with Sternberg. Perhaps the way forward really is the way back—or the way back is the way forward.

But it should be emphasized, as Sternberg says in an *ID the Future* podcast, "I'm not making an occultic argument. It sounds spooky, perhaps, but what I'm arguing is that we can know this on logical and mathematical grounds. It doesn't mean that we can know, say, the source of the 'extra-ontogenetic information.'" (Ontogeny is the process by which the organism takes shape, from conception on, conventionally understood to derive from DNA, the physical genome.) He continues, "It doesn't mean that we know that there's some kind of spooky action at a distance that's taking place. I'm not saying that at all. But I am saying we can figure out, using the existing logic and mathematics that we have today, that some arguments such as those proffered by a standard materialist are simply not true and they cannot be true."

15. Footnotes to Plato

It's all in Plato, all in Plato: bless me,
what do they teach them at these schools?
—C. S. Lewis, *The Last Battle*

THIS BOOK BEGAN VERY MUCH *IN MEDIAS RES*. ALTHOUGH RICHARD Sternberg does bring fresh evidence and a fresh mathematical rigor to bear on the matter, the argument to an immaterial genome is by no means a new one. We have touched on this already, but the subject merits a more sustained look. It's time to travel backward in time, more than two millennia.

In the beginning, there was the Platonic Academy, the academy's eponymous philosopher, and the mostly upper-class men who gathered around him. The Academy was not a university. It likely had no library to begin with; and the students, who included the philosopher Aristotle, are thought to have numbered well under a hundred. It's not certain that Plato himself even taught there, or if he did, what he taught. What's left of the modern site in Athens looks like a patch of dirt in a grove of pine trees with a few wall foundations still to be seen. There are small-town grocers, from back in the day when there were small-town grocers, with more impressive memorials. And yet it is this ancient Greek philosopher who was singled out by the Cambridge philosopher Alfred North Whitehead (1861–1947) when he described the Western philosophical tradition as "a series of footnotes to Plato."[1]

Plato's dialogues, with Socrates usually in the leading role, are wide-ranging conversations, loosely structured, and frequently literary masterpieces. Plato's theory of forms was developed in the dialogue

Phaedo, arguing that beyond our world of physical objects there exists a world of intellectual concepts that inform, or form, those objects. In the *Timaeus* the idea is redolent of contemporary intelligent design theory, with some remarkably modern-sounding features.

Thus, ID has roots stretching back to ancient Greece. So too, as briefly noted previously, does modern evolutionary theory. Men have been arguing for millennia about whether the cosmos reflects the design of an intelligent mind transcending matter, or not. The science writer Daniel Witt has cheekily attributed the "oldest anti-intelligent-design argument"[2] to Epicurus (341–270 BC), a philosopher from the Greek island of Samos, whose own words are mostly lost but who found an eloquent spokesman in a Roman poet and thinker, Lucretius (about 99–55 BC), in his poem *On the Nature of Things.*

Witt summarizes the Epicurean view—but before getting to that, indulge me in mentioning that I've been to Samos. On a college-age visit to Greece, I took a ferry there, across the Aegean Sea with a lot of tanned and good-looking European vacationers. Samos is off today's Turkish coast. On my first night on the island, I couldn't find lodgings and so I slept sitting up in a shelter on the beach. Perhaps Epicurus bathed there—an idea that, I was amused to find recently, I'm not the first to have thought of. A shop on the internet markets items under the business name, The PhiloSurfer Club. It sells coffee mugs and sports towels illustrated with what it calls an "acclaimed design," "Epicurus on the Beach as Surfer in Picasso Style." Get yours now before they sell out. As Richard Sternberg would say, "Just an aside."

Witt, as I was saying, summarizes Epicurus: "Everything that exists was made not by intelligent design, but rather by the random arrangement and rearrangement of atoms. Since the universe is infinite, there are enough opportunities for every possible arrangement of atoms to occur eventually, even the most unlikely. Our world, and the life on it, is one of those unlikely (but eventually inevitable) arrangements."[3]

Plato was of a different mind. That he was an ID proponent ahead of his time is not Richard Sternberg's whimsical notion or my own.

Andrew Gregory is Reader (Professor) in the history of science at University College London. He contributed the introduction to a translation of the *Timaeus* published by Oxford University Press. Gregory writes that its account of "reason and necessity" offers, among other things, an explanation of the "anatomy and physiology of the human body, underpinned by the principles of intelligent design."[4] He is not an admirer of such thinking. The dialogue as a whole, he says, "is infamous for its teleological approach to the explanation of nature"[5]—teleology being another way of saying purposeful design.

Other Greeks—Empedocles, Leucippus, Democritus—advanced a theory very similar to modern proposals of a multiverse, or many worlds, an invention to counter evidence of fine-tuning and design in cosmology. As Gregory summarizes the ancient version of that, "Many of these worlds are generated, all different from one another, and our cosmos is just one of them. It is not in any way designed." As Gregory further summarizes, Plato criticizes this view: "There is one and only one cosmos, which has been designed."[6] Turning to biology, "There are single, fixed species which have been designed."[7]

Gregory feels that Plato's teleology has been superseded, as "the multiplicity-of-accidents view has won out in the form of the theory of evolution."[8] Of course, that is just what modern ID theorists contest. In cosmology, writes Gregory, the "modern question is slightly different from the ancient one"—notice, only "slightly different"!—"in that it asks why the values of certain fundamental constants (such as the speed of light, or the value of the gravitational constant) have values set within the extremely tight limits which allow for the generation of planets and life."[9] In Plato's thinking, a god, the Demiurge, sets "order explicitly by using mathematics, geometry, and harmony." He "looks to an unchanging original," the forms, and "copies" it. The Demiurge's work as a geometer reflects teleology: "It is a matter of intelligent choice and design."[10]

Turning to the dialogue and what Plato himself has to say, we encounter a beguiling mix of the ancient and unfamiliar (and even bizarre), the timeless, and the seemingly very modern.[11] *Timaeus* is not exactly a dialogue, since once Socrates, the first speaker, prompts the title character to display his ideas, Timaeus never really stops, and nothing is said by the philosopher in response to his friend's lecture. It just ends.

The thinking that Plato sketches in the work is an *eikōs logos*, a "likely account," a term that Sternberg could adopt as his own. Anything that begins to be must have a cause, in this case the "beautiful" universe, crafted by that divine being who is "good," consulting an "eternal model."[12] The universe on this view is singular, to the exclusion of an infinite number of universes. The model is eternal, but the universe, and time itself, which was created with the universe, may or may not come to an end.

The "intelligent craftsmanship," a striking phrase (see 46e and 48a of the *Timaeus*), involves primary and secondary causes. The cosmos for Plato, as Gregory puts it in a note, is "designed to support life."[13]

Biologist Michael Denton develops this perspective in *Nature's Destiny* (1998) and further in his "Privileged Species" series of books.[14] He speaks of "remarkable examples of elegance and parsimony in the *fitness of the properties of matter for life*. These are not widely appreciated but they provide for me personally some of the most compelling evidence for intelligent design in the natural world."[15]

The idea has been recognized by others. For instance, Nobel Prize-winning UC Berkeley physicist Charles Townes (1915–2015) was not in the intelligent design community, and yet he said, "Intelligent design, as one sees it from a scientific point of view, seems to be quite real. This is a very special universe: it's remarkable that it came out just this way. If the laws of physics weren't just the way they are, we couldn't be here at all."[16] And Alfred Russel Wallace, the co-founder of the theory of evolution by natural selection, concluded similarly in his 1911 book *The World of Life: A Manifestation of Creative Power, Directive Mind, and Ultimate Purpose*.

The modern scientific myth of the "multiverse" is largely motivated by the desire to evade the implications of "ultimate purpose" manifest in this evident fine-tuning for life. The multiverse hypothesis is, in effect though not intent, an unwitting tribute to the evidence it seeks to deny.

Timaeus goes still further, articulating in nascent form an argument developed by astronomer Guillermo Gonzalez and philosopher Jay Richards in their 2004 book *The Privileged Planet: How Our Place in the Cosmos Is Designed for Discovery*. For Plato, such a design has a moral purpose. Why did the gods give us eyes?

> Let's simply state that the reason and purpose of this gift is as follows: the gods invented and supplied us with vision to enable us to observe the rational revolutions of the heavens and to let them affect the revolutions of thought within ourselves (which are naturally akin to those in the heavens, though ours are turbulent while they are calm). That is, the gods wanted us to make a close study of the circular motions of the heavens, gain the ability to calculate them correctly in accordance with their nature, assimilate ours to the perfect evenness of the gods, and so stabilize the wandering revolution within us.[17]

The intelligent design of human physiology, in short, was to facilitate scientific discovery.

Timaeus depicts two classes of objects: the perfect and eternal (the forms) and a second class of perceptible objects that share their names and resemble them but are copies. (Then there is space, where the creation takes place.) These classes are matched with two sources of causation, the divine and the necessary, which need to be distinguished. "But our concern with divine causes should lead us not to ignore necessary causes either,"[18] Timaeus says, in a way that is roughly similar to the distinction modern ID thinkers make between design and natural law at work in various observable phenomena. In the title of his 2007 book, biochemist Michael Behe calls the line between what natural biological processes can do without design, and what they can't, "the edge of evolution."

Yet another modern ID scientist, bioengineer Stuart Burgess, argues that what evolutionists cite as evidence for the "poor design" of the human body (a bum knee is a commonly cited example) is typically a result of misuse of the body, with the knee, used and exercised as it was intended, representing a remarkable work of engineering—not the best at everything but instead an example of what engineers refer to as constrained optimization.[19] That might make what Timaeus calls "regimen," a change in how we use our body, a likely remedy over drugs and surgery, at least if the problem is caught early enough.

The main takeaway from Plato's thought, however, is that life and the universe were designed, and with human beings foremost in mind. And the source of the design is not in our world, but outside it, in an eternal dimension.

The thesis of the *Timaeus* was tackled, and criticized, by that product of the Platonic Academy, Aristotle. As regards evolution, the Aristotelian legacy is today fiercely fought over, by Catholic philosophers and theologians in particular. This is understandable because Aristotle is not only central to Catholic theology, through his influence on St. Thomas Aquinas, but also because he is a much denser and more difficult writer than Plato. There are no "characters," drama, or witty dialogue as in Plato. Aristotle's work is thought to consist of what were lecture notes, and it reads that way.

As both a Catholic and an evolutionary biologist, Sternberg has written about the Aristotelian controversy. It is a bit like a game of capture the flag: Which side gets to claim Aristotle or Aquinas, that of unguided evolution or intelligent design? Jews have argued similarly about the legacy of Maimonides. Of course, the interest in these great thinkers is not just in pursuit of claiming the revered figure for one's team; it's also because, as we've seen, great thinkers have been arguing about intelligent design versus evolution for thousands of years, and it's not unlikely that they have considered certain questions at considerable depth, even as their empirical or experimental methods left much to be desired.

The problem posed by Sternberg is whether thinking about an immaterial source of information that shapes life ends with Plato (which would be a poor advertisement for it), or whether it continues with Aristotle (which would be more favorable to Sternberg's way of reasoning). This is not a question of interest to Catholics alone, but Sternberg happens to have addressed it in the anthology *God's Grandeur: The Catholic Case for Intelligent Design*. Sternberg argues against those Thomists who wish to see in Aristotle a foundation for materialist evolutionary theories, and who regard Thomas Aquinas's highly Aristotelian natural philosophy as physicalist.

The chapter Sternberg contributed is titled "Why Aristotle Favors Intelligent Design and Not Physicalist Thomism."[20] In it he uses an amusing image, comparing the work of those physicalist Thomists to the work Thomas Jefferson did on the Bible. The so-called *Jefferson Bible* is the four Gospels at the beginning of the New Testament with the miraculous material excised, leaving a "rational" or purely material narrative of the life of Jesus in its place. So too, the evolutionary Thomists disregard everything in Thomas or Aristotle that seems to go against their preferred paradigm.

This subject can feel to outsiders a bit like counting angels as they dance on the head of a pin. Given Sternberg's up-to-date empirical scientific evidence for an immaterial genome, it may even strike readers sympathetic to his argument as largely beside the point. But for those interested in the history of ideas, and for those who want their scientific theories to be not just empirically but also philosophically well grounded, this vein of inquiry will retain a measure of interest. A key issue here is whether Aristotle inherited from Plato an idea of forms as immaterial entities or whether he saw form as inextricably linked with the material world. In other words, was Aristotle a Platonist? Citing the scholarship of University of Toronto philosopher Lloyd Gerson, Sternberg argues that he was, with Aristotle's *Generation of Animals* and his *Metaphysics* echoing Plato's *Phaedo* and *Timaeus*. For both philosophers, form comes before matter, since it is the form, not the matter, which determines the nature of a thing.

In the context of generating life, Aristotle uses for "form" the term *eidos*—hard to distinguish from another term, *psuche*, or the "soul" of the body, which is also its form. When a body changes, as in development from the embryo, it undergoes *kinesis*, or "movement."

In *Generation of Animals*, Aristotle (who of course knew nothing of DNA or of genetic inheritance) discusses the nature of semen, and of "male" and "female," which are not simply genders in the sense we use those terms but principles: "For Aristotle, any construal that something material has to pass from the *form-giver* to *that which is informed* is rejected, and staunchly so," writes Sternberg.[21] Aristotle "presents us with a host of non-physical organizing principles. Here is a notable one: When we open Aristotle to inquire into the 'the physical part of the seed,' we are immediately confronted by its function as the 'vehicle' for an immaterial 'form-principle.'"[22]

Aristotle, like Plato, rejected any role for chance in shaping life. To Aristotle, continues Sternberg, it was plain that "chance acting over extended periods of time could never do anything creative. It will primarily corrupt artifacts and break them apart"[23]—a point emphasized by philosopher of science Stephen Meyer in his books, where he argues that as with computer code, accidents in the genome can be neutral in effect, or they can cause disfunction. They might even, as Michael Behe argues, devolve the organism in a way that provides an advantage to it. But what they can't do, contrary to Darwinian theory, is build up novel, functional form.

The philosophical terms are daunting, but they simply mean that Aristotle followed Plato in seeing the "form" of an animal as issuing from an immaterial source, rather than being conveyed by the semen or seed understood as a purely physical entity. That is why, Sternberg concludes, Aristotle and his follower Thomas Aquinas are a much better match with ID than they are with physicalist theories. Here is Sternberg in his own words: "The argument that Aristotelianism and materialism/physicalism are somehow all of a piece is mistaken. Rather, Aristotle by analogy shows that just as design and agency are necessary in every craft, so too formal and final causes are operative in the reproduction of living things, not just efficient and material causes. In this respect, Aristotle strongly favors intelligent design over any physicalist Thomism."[24]

Sternberg goes further still, to insist that at the very foundation of Western philosophy, we find a firm grounding for the idea of the immaterial genome.

16. On the Origin
of Whales

I F ANIMALS REFLECT DIFFERENT PLATONIC FORMS, WHAT ARE THE implications for evolution and the history of life? It would seem that a marriage between Plato and Darwin would require in the Platonic realm a veritable infinitude of intermediate forms of creatures that once lived on Earth—the opposite of what Plato had in mind.

Scientific hypotheses are rarely if ever evaluated in a vacuum. Typically, there are competing hypotheses, with investigators seeking to make an inference to the best proposed explanation on the table. Stephen Meyer treats this method of reasoning in, among other places, his book *Return of the God Hypothesis*.[1] Sternberg's theory of an immaterial genome, which involves a further inference to an immaterial designing agency, is in competition with an evolutionary paradigm that is materialistic and reductionistic. To assess how these competing approaches stack up against each other, let's turn to a question that the specialist in biology as well as any human with an ounce of childlike wonder left in his body will find of interest, namely the question of the origin of whales.

The title of Hans Thewissen and Sunil Bajpai's 2001 article in the journal *BioScience* struck a triumphalist note: "Whale Origins as a Poster Child for Macroevolution." They confidently informed readers, "The last two decades have witnessed an explosive growth in the number of fossils documenting the origins of Cetacea (whales, dolphins, and porpoises). An excellent morphological series of transitional cetaceans is now available to document the transition from land to sea."[2]

The general idea goes back at least as far as Darwin. In the first edition of *The Origin of Species*, he proposed a land mammal as ancestral to whales—in this case, a bear. "I can see no difficulty in a race of bears being rendered, by natural selection, more and more aquatic in their structure and habits, with larger and larger mouths, till a creature was produced as monstrous as a whale."[3] He was ridiculed for this vague just-so story and informed by at least one scientist friend that the idea gave him a good laugh. He cut the proposal from subsequent editions.[4]

The authors of the *BioScience* article, though, offered comfort to the father of modern evolutionary theory: "Although Darwin didn't have the details right—bears did not evolve into whales—his basic point was correct: We can now show that whales are in fact hoofed mammals that took to sea."[5]

Not so fast.

What had happened in the couple of decades before 2001 was the discovery of a handful of fossils that could be construed as so-called missing links leading from a land ancestor, *Pakicetus* (which resembled a greyhound or wolf) to a whale. The problem in short is this (we'll do the long version in just a moment): In Darwinian terms, the transition from *Pakicetus* to whale would have required not a mere handful of transitional stages but thousands or even millions, so many are the required changes—both genetic and morphological. So finding a few possible (though by no means certain) transitional forms in the fossil record is—to put it mildly—underwhelming. Where are all the missing links, and how could the blind process of evolution by random mutation and natural selection have navigated a pathway from *Pakicetus* to whale in the time allowed by the fossil record—or indeed in any length of time available on planet Earth?

Even granting common descent as I do (and it should be noted that some intelligent design scientists indeed entertain common descent, albeit an intelligently guided form), the speed and directionality of the wildly complex transition from land animal to marine animal still suggest a source of purpose, a movement of striving towards a vision from outside nature itself. Proponents of intelligent design see no objection to this explanation. "An intelligence could have planned to make fully aquatic mammals and designed these features to actualize the plan," as Sternberg's biologist colleague Jonathan Wells wrote.[6] Sternberg differs from other scientists in the ID community in using language—an "extrinsic source of information," and an "informational realm that is outside space and time"—that sounds and is Platonic.

In any case, the problems with whale evolution are so serious that Sternberg argues the origin of whales poses insurmountable problems for neo-Darwinian theory.[7] We could call the event a poster child not for macroevolution but for intelligent design, and perhaps for the immaterial genome.

Sternberg sketches the neo-Darwinian whale story in what he calls "cartoon" form. "You start from a four-legged ancestor that's fully terrestrial," he explains. "It has offspring that begin, say, hunting along margins of the seashore. Its offspring become more amphibious. You have loss of the hind limbs. You have the transformation of the forelimbs into flippers. You have the emergence of a tail fluke. You have offspring, as this is taking place, becoming ever more torpedo-shaped, until finally, you have a more or less modern whale that's completing its life cycle in a fully marine environment."

The purported transitional forms would be "so-called walking whales, or proto-whales." The transition from them would perhaps include *Georgiacetus*, "which shows reduction of the hind limbs, but no tail flukes," and eventually reach truly aquatic beasts, such as *Dorudon* or the sea-serpent-like *Basilosaurus*. And this entire transition from land animal to whale would have to be an event of remarkable abruptness. How abrupt? Sternberg's analysis, in line with other research on the subject, leads him to a conclusion of no more than eleven million years.

To most of us, that sounds like a very long time. For humans, it's about a thousand times as long as it has been since the dawn of agriculture and the neolithic. And yet it's the blink of an eye on geological timescales, and a short enough period that even some evolutionists are uncomfortable with it, given all the morphological and genetic changes required to evolve a whale (changes that we'll enumerate momentarily).

Sternberg sees three time-related problems for the whale-evolution scenario: "One is that we're talking about organisms that have long generation times. On average, we'll say five years. So they're not like bacteria, viruses, or fruit flies or anything like that." Other creatures, like bacteria or fruit flies, are favorites in lab experiments because they reproduce so rapidly and thus you get so many of them. The generational turnover for whales, being so much slower, significantly limits the number of generations, and thus the number of possible "tries" for beneficial genetic mutations across the generations in a given window of geological time.

The second problem, says Sternberg, "is that mammals in general do not have large reproductive population sizes."[8] By comparison, there are roughly a billion trillion malaria parasite cells in human bodies around the world at this moment. Each has an average generational turnover of around one to three days. So we are talking about hundreds of billions of trillions of "baby" parasites every year, each affording a chance at some evolutionarily interesting mutation.

But whales? Back when they were supposedly evolving, "We're talking about breeding populations that may have numbered in the thousands or tens of thousands or perhaps more, but likely not millions or billions." This, combined with their relatively stodgy rate of generational turnover, greatly constrains "the number of genetic changes that could have accrued over time." In short, there weren't enough new proto-whales for nature to experiment on.

A third issue Sternberg underscores is that when you consider the characteristics of the creatures that need to be transformed in the span of just a few million years, "and you begin doing calculations—and this is something that I've been doing with some colleagues—what one finds is that once you get beyond, say, five or six or seven changes that had to occur not necessarily simultaneously, but changes that had to build to an adaptation, it becomes prohibitive. There simply was not enough time for that to occur."

And as one might guess, the number of changes required to transition from *Pakicetus* to whale is vastly greater than six or seven. Sternberg offers an illustration. Say you have a Volkswagen Beetle and decide you want to modify it in order to explore the Mariana Trench, whose bottom is more than six miles deep in the ocean. That would take some very, very extensive re-engineering. By the time you were done getting your car ready for the trip, there would be little if anything left of the poor Volkswagen. Even if you only wanted it to serve as a submersible at much more modest depths, like the sports car/submarine in the James Bond film *The Spy Who Loved Me*, a remarkable vehicle currently owned by Elon Musk,[9] that would take some serious doing.

In the case of whale evolution, the re-engineering would encompass nearly everything. Sternberg lists just a few items:

> Changes to the eyes, changes to hearing, changes to the reproductive system. Modification of the body, so that instead of walking on four limbs, the body adopts a torpedo shape, more or less, that has hydrodynamic properties. It can move very swiftly to hunt prey under the water, so you're talking about changes to the musculature. You are talking about changes to the vertebral column. The origin of a tail fluke, and all of the musculature and neurological systems for coordinating that. Changes in breathing are very important. You have to be able to control breathing in such a way that when diving, the blow hole remains closed, and such that there is the exchange of the exhalation and inhalation once on the surface. You've got to be able to sleep at sea. Mammals sleep. And you've got to do that while most of the body is underwater.

There are more features, including ones that come online very early in embryonic development. How many changes are we talking about? Certainly, it's more than some evolutionists would lead us to imagine. By their lights, all one has to do, Sternberg says, exaggerating for effect, "is shave a cow, cut off its legs, throw it in the water, and it would be able to perform the tasks of a whale."

In pursuit of more realistic estimates, Sternberg cites mathematician David Berlinski, who estimated 50,000 characteristics that would need to be modified. That's "far more right than wrong," according to

Sternberg's own analysis, though he says he might go as low as 5,000 or 10,000. I'm not aware of Darwinists making calculations like this, and perhaps there's a reason for their avoiding the subject, as it would likely damage their cause.

Some have suggested that morphological changes like the reduction of the hind limbs of the proto-whale could have been accomplished through minute micro-evolutionary genetic changes to those limbs and to relevant anatomical features like the reproductive organs. (The method of sexual intercourse would need to be amended.) Sternberg doubts this proposal. The genetic changes across the population, *just for this one morphological change*, would range from the 10s to the 200s. Does the fossil record allow time for that? No. In the available time for the re-engineering of our proto-whale, using a realistic model, the best you could hope for would be five or ten genetic changes. Thus, it's not possible, given the short time that our whales would have available, for blind evolution to manage *even this one* modest re-engineering task.

That brings Sternberg to the so-called "waiting-times problem," an area of his current research under the ID 3.0 program, where he has worked with paleontologist Günter Bechly, mathematician Ola Hössjer, and fellow biologist Ann Gauger. Sternberg summarizes the problem: "It's a question of, if a particular trait needs two or more mutations—not to appear simultaneously, but just for two changes to coalesce, to come together in an individual—such that an adaptive trait appears, how much time would that take?"

That is, under the premise of blind evolution, how long would it take to get an innovation like the tail fluke, which is anatomically complicated? Even if we are considering only two mutations, the waiting can be extraordinarily long. For our own ancestral hominid population, Sternberg cites the research of Cornell mathematicians Rick Durrett and Deena Schmidt in their 2008 article "Waiting for Two Mutations," published in the journal *Genetics*, which gives an estimate of 216 million years.[10] Let that sink in: 216 million years for just two mutations, when dozens to hundreds to thousands would have been needed to evolve a land mammal into a whale.

Whales, by the same calculus but with different generation times and breeding population sizes from hominids, would require 43 million years for *just those two changes*. Yet the *many* novelties we are talking about that differentiate the walking whale from the fully aquatic whale supposedly arose in only a small fraction of that time. By realistic measures it simply couldn't happen.

None of this is to say that the Darwinian process of random variation and natural selection played no role in the history of life. Sternberg agrees with Hugo de Vries who, in his now famous comment, concluded that natural selection, in its purifying role, does shed light on "the survival of the fittest, but it cannot explain the arrival of the fittest."[11]

Alternative evolutionary models like neutral evolution, invoking unguided genetic drift, also can't help. "What we see in whale evolution is directionality," says Sternberg. "We see a number of coordinated changes that occurred very rapidly. And to explain that on the basis of just genetic drift, random changes that just happened to come together and appear to be directed, that does not seem to comport at all with what we see."

What does supply the direction? The information could be coming from inside the system or outside. It could be intrinsic or extrinsic, or some combination of the two. Those are the jointly exhaustive categories, as logicians say.

Sternberg's old colleague, James Shapiro, rejects intelligent design of any flavor in favor of what he calls natural genetic engineering. Asked about that, Sternberg replies:

> I don't want to put words in anyone else's mouth. But it seems to me that that kind of directionality ultimately collapses to a model that's either panpsychist—where matter and mind are basically one and the same thing—or one where, beginning with the Big Bang, there was some predestination. There was a preordering of things, but at a very fundamental level, at a quantum level or an elemental level. Ultimately it would mean that by the time you get bacteria, protozoa, you have these programmed genetic changes. You have mechanisms in place that can diversify and that can be deployed to become everything that we see in the natural world.

Sternberg isn't saying that this is where Shapiro takes it but rather that this is where Sternberg would take it to try to shore up the framework, by his lights. And as we have seen, that isn't the framework Sternberg has chosen to employ to resolve the mystery of the genome, in no small part because such an imaginative extension of Shapiro's model still leaves unanswered the question of how the genetic changes were "programmed." By what or whom?

It is that question that led Sternberg to part ways with the natural genetic engineering model, both Shapiro's more modest instantiation and any model building from it. The "pre-programmer," Sternberg says, can only have been outside the material existence that came to be in the moment of the Big Bang.

Sternberg thus finds that when you eliminate weird forms of naturalistic mysticism such as panpsychism, or contentless games with words like "emergence," you are left with solutions to the problem of directionality that cannot be reconciled with Shapiro's naturalistic model.

What is left to adequately explain the directionality in whale evolution? For Sternberg, it is his "Platonic-Aristotelian" understanding of nature. "I think it's a full-blooded view of the natural world," he says. "It takes each species as being what it is, something that has a paradigm, a model, a verity of its own. And yes, species go extinct; yes, species appear."

How do they appear? "There was a preprogrammed source for that information," he says. "But this is not something we're going to find by just looking at the material elements, by looking at the periodic table or something like that. It's something that we're going to have to find in an extrinsic source, and that extrinsic source is the basis of the rational structure of the universe."

A whale, a dolphin, or any organism points to this "underlying template to the universe," he concludes. "Each and every species has to be taken for what it is. It's of intrinsic value."

Sternberg affirms an argument common among intelligent design theorists, such as Stephen Meyer details in his books and in the article that got Sternberg in so much career trouble, that the materialist/naturalist models on offer fail to account for the evidence from the natural world. "And that," says Sternberg, "is the reason why I would say alternatives such as intelligent design cannot be simply pushed to the side."

Where exactly Sternberg's thinking can be located on the map of ID theory is an interesting question. Meyer, for one, is not an Aristotelian. While grateful for the tradition of Western philosophy with all its gifts, he criticizes the ancient Greek thinkers for relying too much on armchair theorizing and not enough on looking directly at nature. The scientific revolution, he says, happened when scientists inspired by Judeo-Christian tradition sought to understand God's thought by humbly studying His work. In *Return of the God Hypothesis*, Meyer cites a pioneering early scientist who broke with the more deductive approach of the ancient Greek philosophers. "As Robert Boyle (1627–1691), one of the most important figures of the scientific revolution and the founder of modern chemistry, explained, the job of the natural philosopher was not to ask what God must have done, but what God actually did. Boyle argued that God's freedom required an empirical and observational approach, not just a deductive one. Scientists needed to look, and to find out."[12]

Another leading ID advocate, neuroscientist Michael Egnor, locates himself in the Aristotelian tradition, as mediated by Thomas Aquinas. Egnor writes, "There is deep beauty and encompassing rationality to the Aristotelian-Thomist way of understanding the world. It can be said that Aristotle was the last man to know everything that could be known in his time, and that Aquinas was the last great systematic philosopher—the last philosopher/theologian to put together a coherent system for understanding all of reality."[13]

Sternberg also reminds me of another ID scientist with an Aristotelian bent, the Australian biologist Michael Denton, who argues for "a modern restatement of Aristotle's notion of forms as active agencies in nature, responsible for the generation of the particular set of biological forms or types manifest in life on Earth."[14]

The ID thinker whose views most closely align with Sternberg's may be German paleontologist Günter Bechly, who all but fully avowed the title of Neoplatonist. I will return to him shortly.

Beauty, rationality, intrinsic value: these are hallmarks of design thinking, and these are what ID proponents, whether Platonist, Aristotelian, or otherwise, find reflected in the natural world.

17. The Interpretation
of Dreams

On a visit to Rome in 2007, where he lectured at the Pontifical Gregorian University, Richard Sternberg had tired of Italian food. He decided, on his final night in the ancient city, to dine at a Chinese restaurant. Afterward, around 3 a.m., he awoke from a vivid dream that would turn out to be life-changing. It kept him up for the rest of the night and into the next day. In the dream, he saw a figure looking like an Orthodox Christian monk, "an archimandrite wearing full schema, a monk's vestments, and holding out a large book, in ornate, bejeweled, and gold and silver casements." The monk displayed to him "intricate diagrams, layers and layers and layers. It looked very kabbalistic, if I may say." The book showed "organisms, parts of the organisms and the tissues and cells." Each, Sternberg felt, held a symbolic meaning, was itself a text. "I was astounded by it," he told me. In the dream, the monk then spoke: "'All this knowledge has been given to men to lead to repentance. Time is short.' And then he closed the book."

What were the organisms? I pictured trilobites, tiger prawns, whales. But Sternberg cannot recall.

For Sternberg, "repentance" here meant not rejecting and giving up sin, as it is conventionally understood, but something closer to the Greek, *metanoia*, literally to change or reattune one's mind and outlook, or the Hebrew *teshuvah*, to turn or return. Says Sternberg, "I had to attune my mind to realizing that there are these deeply logical, symbolic, mathematical layers in the things I was studying. That one had to, so to speak, see beyond the surface. It's like Blake, seeing infinity in a blade of grass, a grain of sand."

The reference is to the poet William Blake in his poem "Auguries of Innocence," where he writes, "Hold Infinity in the palm of your hand /And Eternity in an hour." This is another way of expressing the kabbalistic *tsimtsum*, the contraction of the infinite in the finite, the immaterial in the material, the nature of the genome that Sternberg studies.

Whatever one makes of the dream, its mixing of religious and scientific images prompts a final question about his thesis: "But is it science?"

In the conversation I mentioned in Chapter 1 between Sternberg and Francis Collins, hosted by Charles Colson, one of Collins's criticisms of intelligent design was that it illegitimately crossed boundaries between science and theology. That is the subject of this chapter. Collins himself, in the same conversation, crossed boundaries between science and faith. Because of fused chromosomes in the chimp and human genomes, he explained, the scriptural account of the unique creation of man could not be true. "Look at what the f**k you just did," Dr. Sternberg recalls thinking at the time. (I told you he cursed.) "It's the very thing he had accused ID of doing."

As I was writing this, evolutionary biologist Colin Wright took a poll on X. He posed the following question: "Is Intelligent Design a scientific hypothesis?" Wright is a Fellow at the Manhattan Institute and an advisor for Atheists for Liberty. He has received criticism for defending the scientific objectivity of gender. I thought it could be worth telling him that no less a figure than atheist biologist Richard Dawkins finds the idea of a cosmic designer to be a "scientific hypothesis" albeit, Dawkins thinks, a mistaken one.

In 2024 in New York, Dawkins participated in a moving public dialogue with former New Atheist Ayaan Hirsi Ali, who is also a former Muslim. She had not long before announced her conversion to Christianity from atheism for what she describes as "very subjective" reasons. It was in response to a "personal crisis": "I lived for about a decade with intense depression and anxiety and self-loathing. I hit rock bottom. I went to a place where I actually didn't want to live anymore but wasn't brave enough to take my own life." Faith rather than suicide was her way out of the crisis.

Dawkins answered kindly that belief in a designer is more than a mere subjective response: "You appear to be a theist," he told her. "You appear to believe in some kind of higher power. Now, I think that the hypothesis of theism is the most exciting scientific hypothesis you could possibly hold." Hold that thought in your mind.

Unsurprisingly, Dawkins wasn't giving up his own atheism. But he did further emphasize the point above: "The idea that the universe was actually created by a supernatural intelligence is a dramatic, important idea. If it were true, it would completely change everything we know. We'd be living in a totally different universe. That's a big thing. It's bigger than personal comfort and nice stories and these things. The idea that the universe has lurking beneath it an intelligence or supernatural intelligence that invented the laws of physics, that invented mathematics, is a stupendous idea, if it's true."[1]

What Dawkins is describing here is the design argument from the fine-tuning of the laws of physics and from "the unreasonable effectiveness of mathematics"[2] in science, championed by everyone from intelligent design proponent Stephen Meyer to Nobel-Prize winning scientists such as Arno Penzias and Charles Townes. And although Dawkins rejects the argument as untrue, he frankly concedes that the argument is scientific. This is a remarkable admission, granting what design theorists have long insisted, namely that the theory of intelligent design is every bit as scientific as a materialistic theory such as neo-Darwinism.

The design hypothesis could be wrong, or it could be right. But we should weigh it on its own terms as the scientific hypothesis it is. Thank you to Richard Dawkins for pointing that out.

As philosophers of science have shown, there are multiple, differing, sometimes overlapping criteria for what constitutes a scientific theory, the criteria differing based on the particular scientific discipline in question and on who's doing the scorekeeping. This is why there is no quicker way to drive a philosopher of science to distraction, particularly one with a good grasp of the evolving and differing methodologies of multiple scientific disciplines, than to ask, What is *the* checklist for a properly scientific theory?

Meyer has written extensively on what in the philosophy of science is known as the demarcation issue—that is, the issue of defining the boundary or boundaries between science and non-science. As he shows, whatever conventional criteria one invokes, intelligent design fares just as well as Darwinism, the only exception being the question-begging rule—rightly rejected in various mainstream scientific disciplines—that intelligent causes must not be invoked in a scientific theory. Both theories involve inferences to what occurred in the distant past. Both appeal to a suite of physical evidence. And as Meyer shows, reasoning in the field of intelligent design draws on Darwin's own method of historical scientific reasoning.[3]

Regarding Sternberg's particular design argument, there is also this: Many scientists, rightly or wrongly, carry a bias in favor of scientific investigation and scientific theories focused on natural phenomena that can be observed happening repeatedly in the present. Not for them are the mysteries of ancient natural history; give these scientists the activities of nature in the here and now.[4] The prejudice is understandable. While the historical sciences have made extraordinary breakthroughs in unraveling many mysteries of the past, it is undeniable that they face a daunting challenge not faced by scientists focused on present phenomena: namely, the historical scientist is investigating something that happened in the past and cannot study it at the time it was occurring (for example, the Cambrian explosion).

What does all this have to do with Sternberg? The eerie and unusual nature of his conclusions might lead one to overlook the fact that his subject matter faces no such challenge. If he is right, Sternberg strengthens the case for design by inferring that intelligent agency operates in life at this very moment.[5] His theory explains phenomena that occur repeatedly *in the present*—in every cell, in every developing embryo.

Yes, Sternberg's immaterial genome hypothesis invokes something immaterial, so on the surface it would seem to fail this criterion.

We can't actually *see* this immaterial input. But as with various iconic theories in physics, it invokes for evidence observable effects that repeatedly occur in the present. One cannot observe the law of gravity as an entity in itself. But one can observe its effects repeatedly in the present. In the same way, one can observe the effects of an immaterial genome that veritably shouts teleology, busy as an entire hive of bees working in marvelous concert, and doing marvelous, life-giving things right under our noses that defy the material-genome paradigm.

In contrast, modern evolutionary theory's explanation for the Cambrian explosion posits a causal story (gradual evolution via random mutations and natural selection) that, until a time machine is invented, we have no means of directly observing. In this regard the design hypothesis for the Cambrian explosion is in the same boat, a boat crowded with theories in the historical sciences, from design to evolutionary biology to historical geology and all manner of historical sciences in between.[6]

One might object that the immaterial genome and the agency it suggests are not only immaterial but teleological, thereby dooming the immaterial genome hypothesis to the status of being unscientific. But such an objection boils down to question-begging—the materialist ruling out immaterial teleological causation by fiat, replacing the scientific virtue of following the physical evidence where it leads with a game of victory-by-defining-the-rules.

If we set aside such games, we see that arguments for intelligent design can be advanced in terms that any scientist, if being strictly reasonable, would have to recognize as scientific. One virtue of a scientific theory is that it makes testable predictions. Obviously it is better if the predictions prove correct. As we saw, the Darwinian paradigm led the great bulk of the evolutionary biology community to see non-coding DNA as junk, whereas the theory of intelligent design motivated a prediction of function for this so-called "junk DNA." The ID paradigm was proved correct.

In 2024, the Nobel Prize in Physiology or Medicine was awarded to Victor Ambros and Gary Ruvkun for demonstrating that DNA previously supposed to be junk in fact plays a vital role in generating microRNA, indispensable in regulating genes and thus indispensable for life. A Nobel Prize sounds quite scientific. Casey Luskin wrote, "That so-called genetic junk would turn out to be functional was a prediction of intelligent design going back to the 1990s. On that, ID has been vindicated over and over again, now by the Nobel Committee. Our colleagues Richard Sternberg and Bill Dembski were early predictors, as critics of what Jonathan Wells called in a 2011 book *The Myth of Junk DNA.*"[7]

As I have mentioned, Sternberg's major collaborator in his recent work was German paleontologist Günter Bechly. When Bechly died, he was working on a book about the fossil record. I hope his colleagues will have that ready for publication at a future date. But in one of the last recorded lectures we have of him, delivered at Cambridge, England, he gave what may have amounted to a summary of his argument in the planned book.[8] For anyone with questions about the scientific status or rigor of ID, I encourage you to watch that video.

Bechly followed the method of inference to the best explanation, an approach firmly established in modern science. But as he said in an article co-written with Meyer, in a footnote he included there, he, like Sternberg, thought life might represent "Platonic forms… in the mind of an intelligent designer."[9] On his personal website, he explained further:

> Instead of neo-Darwinism, I endorse a modern version of salta-tionism, mutationism, and orthogenesis, based on non-random adaptive macro-mutations (analogous to Schindewolf's and Gold-schmidt's "hopeful monster" hypothesis, more recently endorsed by Rieppel 2017), correlated with the spatiotemporal instantiation of non-material and eternal templates (Platonic forms) that function as attractors ("special transformism" *sensu* Chaberek 2017), and are quasi-"downloaded from the cloud."

The image of forms "downloaded from the cloud" is worth noting. As a modern technological analogy for the immaterial genome, it may be the most accurate I have heard.

Sternberg described to me the 2011 conference in Santa Barbara, California, where he first sketched the Platonist understanding of ID for a group of scholars that included Bechly. That night the German scientist, as with Sternberg in Rome, was disturbed in his rest. He could not sleep at all as he thought about the implications.

Bechly's essay with Meyer, "The Fossil Record and Universal Common Ancestry," has since proved to be a key contribution in the field. It highlights "the many discontinuous or abrupt appearances of new forms of life in the fossil record—a pattern that contradicts the continuous branching tree pattern of biological history postulated by proponents of universal common descent."[10] There, in a *tour de force*, the authors reviewed nineteen fossil explosions in the history of life that testify against Darwinism and support design. After all, according to the Latin motto that Darwin repeated six times in *The Origin of Species*, "*Natura non facit saltum*"—"Life does not take leaps." Darwin understood that on his evolutionary model, we should expect forms to shade over from one to another with extraordinary gradualism—no great leaps in terms of new biological forms. But on Sternberg's Platonic design model, why not? At least it is possible.

Before closing I want to further clarify the logic of contemporary design arguments in biology, and how Sternberg's theory fits into this larger class of arguments. In *Return of the God Hypothesis*, Stephen Meyer rebuts criticisms that intelligent design commits the "God of the gaps" fallacy, or is an "argument from ignorance," and thus cannot claim the mantle of science. In advancing this complaint, according to Meyer, critics mischaracterize the ID argument as boiling down to a single premise leading directly to a conclusion, thus:

Premise: Material causes cannot produce or explain specified information.

Conclusion: Therefore, an intelligent cause produced the specified information in life.[11]

That would obviously be an absurd way to reason. But it is a strawman of the contemporary design argument in biology. As Meyer shows, it leaves out a crucial second premise and oversimplifies the first premise and the conclusion. Using the origin-of-life question to illustrate, Meyer explains that the actual intelligent design argument in biology instead proceeds thus:

Premise One: Despite a thorough search, no materialistic causes have been discovered with the power to produce the large amounts of specified information necessary to produce the first cell.

Premise Two: Intelligent causes have demonstrated the power to produce large amounts of specified information.

Conclusion: Intelligent design constitutes the best, most causally adequate explanation for the origin of the specified information in the cell.[12]

"Most causally adequate" is obviously a comparative statement. There could be a more, and a less, adequate explanation, as far as we know. The context for what Meyer writes is the origin of the first cell, but the same reasoning applies in other contexts, including the explosions of biological novelty catalogued by Meyer and Bechly, and the development of an embryo into a fully formed organism.

The immaterial genome thesis tells us a material source of information alone cannot account for these wonders. That is, the operations of a solely material channel for information in the cell (i.e., DNA plus epigenetic information) are not causally adequate. According to scientific—not religious—reasoning and observation, only an immaterial and teleological source of information, within and beyond the physical cell, is fully adequate.

EPILOGUE:
BEYOND DARKNESS

*There, peeping among the cloud-wrack above a dark tor high up in
the mountains, Sam saw a white star twinkle for a while. The beauty
of it smote his heart, as he looked up out of the forsaken land, and
hope returned to him. For like a shaft, clear and cold, the thought
pierced him that in the end the Shadow was only a small and passing
thing: there was light and high beauty for ever beyond its reach.*
—J. R. R. TOLKIEN, *THE RETURN OF THE KING*

IF YOU HAVE COME THIS FAR WITH ME, YOU CAN NOW BRAG TO YOUR
friends that you have read a book by a Neanderthal author. Or partially Neanderthal. In the course of completing this book I received
the results of the 23andMe ancestry test I mentioned earlier. Among
other findings from 23andMe, I learned that my DNA is under
two percent Neanderthal. Also, the test, disappointingly, showed
0.0 percent Ashkenazi Jewish. Among the vast majority of my fore-
bearers who were from Northwestern Europe, the largest share was
51+ percent Swedish, from Sweden's Götaland, which gave us the
Old English "Geat" hero Beowulf. The test reports a likely Native
American grandparent from between 1690 and 1780 and, intriguingly,
a high likelihood of my being 0.1 percent North African or Western
Asian.[1] What does any of this tell me about who I am? Heredity can
inspire, but as I learned from personal experience, as related in the
introduction, it can also horrify.

Evolution in Darwin's hands does not sound so bad at first. He wrote in the conclusion of *The Origin of Species*: "There is grandeur in this view of life, with its several powers, having been originally breathed into a few forms or into one; and that, whilst this planet has gone cycling on according to the fixed law of gravity, from so simple a beginning endless forms most beautiful and most wonderful have been, and are being, evolved."

But very quickly others saw horrors embedded in Darwin's view of life. "The next generation of biologists were less confident and consoling," observes literary historian Roger Lockhurst in the collection *Late Victorian Gothic Tales*. "Using Darwin's theory, and the many rival biological accounts of development then in circulation, scientists suspected that it was just as possible to *devolve*, to slip back down the evolutionary scale to prior states of development."[2]

English and American Gothic literature uses inheritance as one of its primary themes with which to spook readers. A generation before Darwin unveiled his theory, Edgar Allan Poe grasped what could be terrible about inheritance, making it the centerpiece of his 1839 story "The Fall of the House of Usher." A doomed brother and sister have inherited a malign house, which collapses in on itself at the climax. Post-Darwin, Victorian and later fiction writers saw how a scientific gloss, supplied by Darwinian theory, could make heredity even more horrible. Lockhurst writes, "There are constant anxieties about maternity and birth, about what is inherited from the mother…. There is also inheritance in the more strictly biological sense: what residues of the primitive or the animalistic lurk in the modern body and mind?"[3]

H. G. Wells's *The Time Machine* evokes horror with a devolved race of the distant future, the troglodyte Morlocks (1895). Robert E. Howard (creator of Conan the Barbarian) and Welsh author Arthur Machen both wrote short stories set in their own time that have as a premise the existence of a barbarian, stunted, unevolved humanoid race hiding underground from the modern-day Britons. See Machen's "The White People" (1904) for a chilling example. This is not so much devolution but something related: stagnating evolution. The prominent arch-atheist horror writer H. P. Lovecraft (1890–1937) dwelled on what he saw as the genetic horrors of race mixing, which he extended in tales like "The Shadow over Innsmouth" (1931) to humans concealing evidence in themselves of miscegenation with aliens, the Old Ones, fallen from space. And so on.

There are many variations. Well before the United States and then Nazi Germany experimented with eugenic sterilization or extermination, men of imagination saw that there was a dark side to the material genome. It rendered human beings a product of, in the final analysis, coded beads on a string. As Lockhurst points out, there was always the possibility of devolution, or of human races that are more and less "evolved." Beyond that, there is the nature of the "particulate" inheritance itself. For many of us, "particulate" is a worrisome word, conjuring the particulate matter in the air that medical scientists measure—too small to be detected by the naked eye—as a threat to our health. Inhaling unseen particles, whether the Covid virus or wildfire smoke, is reflexively understood as something to avoid. When we hear about something being particulate, we are right, it seems, to be concerned.

So, will the materialistic vision of the ancient atomists, the Victorian Darwin, and modern-day neo-Darwinists triumph, or will it be the vision of thinkers stretching from Plato through Johannes Kepler (another student of the *Timaeus*), Alfred Russel Wallace, Gregor Mendel, and now to Sternberg? Will our culture's fundamental view of man and nature be shaped by belief in a purely material genome, or by the immaterial genome?

The stakes were stated darkly by the French atheist biologist Jacques Monod in his book *Chance and Necessity* (1970), his title being a reference to the pre-Socratic Democritus with his view that "Everything existing in the Universe is the fruit of chance and of necessity."[4] Monod ends his book with the depressing thought that under his materialist vision, "The ancient covenant is in pieces; man knows at last that he is alone in the universe's unfeeling immensity, out of which he emerged by chance. His destiny is nowhere spelled out, nor is his duty. The kingdom above or the darkness below: it is for him to choose."[5]

So we find that the debate we have been exploring traces quite clearly from Greek antecedents (Democritus, Plato) to modern exponents (Monod, Sternberg). There is indeed a choice here.

As I was writing this, my family had a conversation about a post on the X platform by the right-wing (I wouldn't want to call him conservative) commentator Mike Cernovich. It was just after the 2024 presidential election and candidate Kamala Harris's team had been overseeing her presence on X as it was shutting down its operation. They posted a couple of photos of the group. Cernovich, who has 1.3 million followers on X, shocked me by saying something beyond nasty. "You are all sickly and unhealthy," he typed. "REVOLTING! I am 47 and more vital. This is pathetic. The old men are supposed to fear the young. You are dysgenic and genetic dead ends."[6] The comment was itself a fiction, since whatever Cernovich imagined, though a couple of the young people struck awkward poses, the group otherwise appeared like normal and healthy twentysomethings.

But beyond the absurd public bragging about his own "vitality," it was the references to "dysgenic" and "genetic dead ends" that caught my attention. Such thinking is very much with us today, and it struck me that conceiving of the gene in a partly spiritual, Platonic light may not only be the better science; it would also make genuine nonsense of Cernovich's ugly insult and other talk like it. So it has that going for it.

It has something more as well. Since science and culture are inextricably linked, I believe an immaterial connection between us all, our parents and our children and the rest of life, might add a touch of grace to our troubled, often ugly culture. To revise Darwin's phrase, there is grandeur in this view of life that Richard Sternberg points us to, with an unseen power that makes each life possible. That view was intimated by Plato long ago and would seem to be on its way to vindication by recent scientific discoveries.

Given the foresight, the planning, the care and creativity apparently exercised by this power, it is understandable that some of our greatest thinkers have seen in this cause something personal. Darwin's colleague Alfred Russel Wallace, co-discoverer of the theory of evolution by natural selection, came to just such a conclusion. In a 1910 newspaper interview he speculated about "spirits, angels, gods, what you will; the name is of no importance." Wallace was speaking of evidence of a personal, purposive "control in the lowest cell." His conclusion: "The wonderful activity of cells convinces me that it is guided by intelligence and consciousness."[7] This is just a simpler way of saying what Sternberg says about, for example, the spliceosome.

Moved by the evidence he had found, Wallace went on: "It may not be possible for us to say how the guidance is exercised, and by exactly what powers; but for those who have eyes to see and minds accustomed to reflect, in the minutest cells, in the blood, in the whole earth, and throughout the stellar universe—our own little universe, as one may call it—there is intelligent and conscious direction; in a word, there is Mind."[8]

However one chooses to label this power, it is a creative intelligence that reaches down to us, it would seem, at every moment, touching us, as the poet William Butler Yeats put it in another context, in "the deep heart's core." It reaches us *within* space and time but *from beyond* their confines—with care, with love, as I picture it, a touch that is more beneficent than the gentlest caress.

ENDNOTES

INTRODUCTION

1. Siddhartha Mukherjee, *The Gene: An Intimate History* (New York: Scribner, 2016), 2.

2. Craig Holdrege, "Goethe and the Evolution of Science," *In Context* (Nature Institute) no. 31, Spring 2014, https://www.natureinstitute.org/article/craig -holdrege/goethe-and-the-evolution-of-science.

3. David Klinghoffer, "The Branding of a Heretic," *Wall Street Journal*, January 28, 2005, https://www.wsj.com/articles/SB110687499948738917.

4. David Klinghoffer, "Unintelligent Design," *National Review*, August 16, 2005, https://www.nationalreview.com/2005/08/unintelligent-design-david-klinghoffer/.

CHAPTER 1

1. Stephen C. Meyer and Günter Bechly, "The Fossil Record and Universal Common Ancestry," in *Theistic Evolution: A Scientific, Philosophical, and Theological Critique*, ed. J. P. Moreland et al. (Wheaton, IL: Crossway, 2017), 344.

2. Eugene Koonin, "The Biological Big Bang Model for the Major Transitions in Evolution," *Biology Direct* 2, no. 21 (August 2007), https://link.springer.com /article/10.1186/1745-6150-2-21. Cited by Meyer, "Neo-Darwinism and the Origin of Biological Form and Information," in *Theistic Evolution*, 108.

3. Gerd B. Müller and Stuart A. Newman, eds., *Origination of Organismal Form: Beyond the Gene in Developmental and Evolutionary Biology* (Cambridge, MA: MIT Press, 2003). Cited by Stephen Meyer, "The Origin of Biological Information and the Higher Taxonomic Categories," *Proceedings of the Biological Society of Washington* 117 (2004): 213–239, available at https://www.discovery.org/a/2177/.

4. Meyer, "The Origin of Biological Information."

5. "Researcher Claims Bias by Smithsonian," *Washington Times*, February 13, 2005, https://www.washingtontimes.com/news/2005/feb/13/20050213-121441-8610r/.

6. Francis Collins, *The Language of God: A Scientist Presents Evidence for Belief* (New York: Free Press, 2006), 66, 67.

7. I quoted from Sternberg's OSC complaint in David Klinghoffer, "The Branding of a Heretic," *Wall Street Journal*, January 28, 2005, https://www.wsj.com/articles /SB110687499948738917.

8. James McVay for the US Office of Special Counsel to Richard von Sternberg, August 5, 2005, transcribed at *Richard Sternberg*, https://richardsternberg.com /smithsonian/letter/. For a scanned PDF of the letter, see *Discovery Institute*, https://www.discovery.org/m/2008/02/OSC-Sternberg-preclosure-ltr2.pdf.
9. McVay for the US Office of Special Counsel.
10. McVay for the US Office of Special Counsel.
11. "List of Peer-Reviewed and Mainstream Scientific Publications Supporting Intelligent Design," *Discovery Institute*, last modified May 2024, https://www .discovery.org/m/securepdfs/2024/05/Peer-Reviewed-and-Mainstream-Articles -Page-Update-May-2024_FinalPDF.pdf.

CHAPTER 2

1. "What Is a Gene?," *MedlinePlus*, accessed February 8, 2025, https://medlineplus. gov/genetics/understanding/basics/gene/.
2. Siddhartha Mukherjee, *The Gene: An Intimate History* (New York: Scribner, 2016), 11.
3. Mukherjee, *The Gene*, 24. Emphasis in original.
4. Charles Darwin to Asa Gray, September 5, 1857, *Darwin Correspondence Project*, Letter no DCP-LETT-2136, University of Cambridge, https://www.darwinproject .ac.uk/letter?docId=letters/DCP-LETT-2136.xml. Quoted by Richard Owen, review of *Origin* and other works, *Edinburgh Review* 111 (1860): 530, https://darwin -online.org.uk/content/frameset?itemID=A30&viewtype=text&pageseq=1.
5. Weismann's experiment was primarily aimed at Lamarckian inheritance, but Darwin did include Lamarckian ideas in his theory, especially in the 6th edition of *The Origin of Species*.
6. Mukherjee, *The Gene*, 63. Emphasis in original.
7. Mukherjee, *The Gene*, 71.
8. Mukherjee, *The Gene*, 95.
9. Mukherjee, *The Gene*, 115.
10. Mukherjee, *The Gene*, 125.
11. Mukherjee, *The Gene*, 178.
12. "How to Build a Worm," Discovery Science, *YouTube*, April 15, 2015, video, 9:39, https://youtu.be/QDQ0NJQ_z3U?si=5Tfhr7wE7YCirhGO.
13. Mukherjee, *The Gene*, 191.

CHAPTER 3

1. Richard Sternberg, "How My Views on Evolution Evolved," *Discovery Institute*, January 2008, https://www.discovery.org/m/2008/01/sternintellbio08.pdf.
2. Sternberg, "How My Views on Evolution Evolved."
3. Barbara McClintock, "The Significance of Responses of the Genome to Challenge," *The Nobel Foundation*, accessed January 21, 2025, https://www.nobelprize .org/uploads/2018/06/mcclintock-lecture.pdf.

4. See biologist Jonathan Wells, *The Myth of Junk DNA* (Seattle, WA: Discovery Institute Press, 2011) and more recent writing on the subject by Sternberg, Casey Luskin, and Jonathan McClatchie at *Evolution News and Science Today*.

5. Sternberg, "How My Views on Evolution Evolved."

6. Sternberg, "How My Views on Evolution Evolved."

7. Sternberg explains the distinction between soft structuralism and hard structuralism as follows: "A 'soft' structuralism is one where various 'self-organizational' processes are accorded roles that are of almost the same significance as that of 'genetic regulation.' (Stephen J. Gould would be a member of such a school I think.) A harder structuralism, then, is one where sundry physical and 'self-organizational' forces are taken to be of greater significance than inherited parameter sets. That said, a hard structuralism is one where the formal generative rules are sought that would allow not only an understanding of the morphogenesis of an organismal form, but also the possible variations of that said form. (Goethe, Driesch, D'Arcy Thompson, and Thom would be members of such a school.) In its hardest version one would seek an organismal 'Schrödinger equation' that would circumscribe, a priori, all permissible evolutions of that species or type."

8. Sternberg, "How My Views on Evolution Evolved."

9. Sternberg, "How My Views on Evolution Evolved."

CHAPTER 4

1. Benjamin Franklin, *Benjamin Franklin's Autobiography and Selections from His Other Writings* (New York: Modern Library), 21. Emphasis in original.

2. Peter Tompa and George D. Rose, "The Levinthal Paradox of the Interactome," *Protein Science* 20, no. 12 (October 10, 2011), 2074–2079.

3. The arguments in the sources listed here are beyond the scope of this book. I asked Sternberg about Shannon's 10th theorem. How would he put the matter in a nutshell? He replied in an email, "Shannon proved that even with the least reliable channels of data, one can create an arbitrarily precise level of signal transmission. This holds for any channel that has any error rate (save for exactly 50 percent per bit). Say that you have a channel that sends the correct bits of data only 51 percent of the time—one that sends the correct data at only a slightly higher rate than it does those which are errors—you can still pass on data such that *only one bit* out of a hundred, thousand, million, billion, trillion, et cetera *is wrong*. This is achieved through *redundancy*." See C. E. Shannon, "A Mathematical Theory of Communication," *The Bell System Technical Journal* 27 (1948), 379–423. For more on Rosen, see Chapter 7.

4. Richard Sternberg, "How My Views on Evolution Evolved," *Discovery Institute*, January 2008, https://www.discovery.org/m/2008/01/sternintellbio08.pdf.

5. Michael Levin, "Ingressing Minds: Causal Patterns Beyond Genetics and Environment in Natural, Synthetic, and Hybrid Embodiments," *PsyArXiv* (February 7, 2025), https://osf.io/preprints/psyarxiv/5g2xj_v3. Invoking Plato and Pythagoras in a biological context in 2025 is, of course, not the norm. The appearance of Levin's paper struck me as a remarkable instance of synchronicity. For bonus

points, the final citation at the end of the paper is to Carl Jung's book *Synchronicity*. See also an interview with University of Zürich evolutionary biologist Andreas Wagner, also invoking Plato, referenced in Daniel Witt, "Is Evolution's 'Third Way' Natural? (And Are We Allowed to Reference It?)," *Evolution News and Science Today*, March 25, 2024, https://evolutionnews.org/2024/03/is-evolutions -third-way-natural-and-are-we-allowed-to-reference-it/.

6. Sternberg, "How My Views on Evolution Evolved."

7. Collins's natural theology is not consistently deistic; it is married uneasily to his theistic Christian faith.

8. Sternberg, "How My Views on Evolution Evolved."

9. See, for example, Casey Luskin, "Pseudogenes Aren't Nonfunctional Relics that Refute Intelligent Design," *Evolution News and Science Today*, September 9, 2021, https://evolutionnews.org/2021/09/pseudogenes-arent-nonfunctional-relics -that-refute-intelligent-design/.

CHAPTER 5

1. The direct quotations that follow are from Sternberg's Center for Science and Culture Summer Seminar lecture in 2020, "What Is a Gene? Not a Particle, but a Process: How the Gene Is a Multilevel Mediator of Information that Currently Lacks a Material Description." He has been teaching in the Summer Seminar since 2007, but 2020 was of course the year of Covid lockdowns, so he taught on Zoom and the lecture was conveniently captured online.

2. The immaterial genome necessarily remains unsequenced. Also, see Michael Marshall, "Why the Human Genome Was Never Completed," *BBC*, February 12, 2023, https://www.bbc.com/future/article/20230210-the-man-whose-genome -you-can-read-end-to-end.

3. Unless otherwise stated, the shorthand term "evolutionist" refers to evolutionists whose evolutionary models are restricted to purely materialistic causes, whether appealing to natural selection, genetic drift, or other wholly mindless material processes.

4. Richard Dawkins, *The Greatest Show on Earth: The Evidence for Evolution* (New York: Free Press, 2009), 333.

5. "Jonathan Sacks and Richard Dawkins at BBC RE:Think Festival 12 September 2012," Brian Sacks, *YouTube*, September 14, 2012, video, 1:05:28, https://youtu.be /roFdPHdhgKQ?si=rUjDIowopnM-PlNp.

6. Francis S. Collins, *The Language of God: A Scientist Presents Evidence for Belief* (New York: Free Press, 2006), 136.

7. Carl Zimmer, "Is Most of Our DNA Garbage?," *New York Times*, March 5, 2015, https://www.nytimes.com/2015/03/08/magazine/is-most-of-our-dna-garbage .html/.

8. See Aristotle's *Generation of Animals*, trans. A. L. Peck (Cambridge, MA: Harvard University Press, 1943), Book I, section 722a-b, 57, https://archive.org /details/generationofanim00arisuoft/page/56/mode/2up?q=syllable.

9. Sternberg cites Joseph Needham, *A History of Embryology*, 2nd ed. (Cambridge: Cambridge University Press, 1959), 53–54: "If I have devoted such ample space to an account of Aristotle's contributions to embryology, it is, firstly, because they are actually greater in number than those of any other individual embryologist, and secondly, because they had so profound an influence upon the following twenty centuries."

CHAPTER 6

1. Gregor Mendel, "*Versuche über Pflanzenhybriden*," *Verhandlungen des naturforschenden Vereines in Brünn*, Bd. IV für das Jahr, 1865, *Abhandlungen*: 3–47. For an English translation see "Experiments in Plant Hybridization," trans. William Bateson and Roger Blumberg, *The Electronic Scholarly Pursuit Project*, 1996, http://www.esp.org/foundations/genetics/classical/gm-65.pdf.

2. Thanks goes to the late Dr. Bechly for help with the translation.

3. Maxwell considered the question of what an atom was. Sternberg says Maxwell knew that the "idea of an atom as this little hard unit like a marble, but incredibly small, that old, materialistic notion, was simply wrong. And he said if that's wrong, then trying to build higher order concepts on it was also wrong, and that's exactly what Darwin, Haeckel, and the others were doing." Inheritance was not a matter of marbles, or beads on a string. It couldn't be if the foundation of matter was the atom as Maxwell understood it.

4. W. Bateson and Miss E. R. Saunders, *Reports to the Evolution Committee: Report I* (London: Royal Society, 1902), 147.

5. See the previous chapter.

6. T. H. Morgan et al., *The Mechanism of Mendelian Heredity* (New York: Henry Holt and Company, 1915), viii. Emphasis added.

7. Daniel Witt, "Is Vitalism Making a Comeback?," *Evolution News and Science Today*, May 21, 2024, https://evolutionnews.org/2024/05/is-vitalism-making-a-comeback/.

CHAPTER 7

1. Erwin Schrödinger, *What Is Life?* [1944] (Cambridge, UK: Cambridge University Press, 2023), 22.

2. Joan Bagaria, "Set Theory," *Stanford Encyclopedia of Philosophy*, eds. Edward N. Zalta and Uri Nodelman (Spring 2023), https://plato.stanford.edu/archives/spr2023/entries/set-theory/.

3. Robert Rosen, *Life Itself: A Comprehensive Inquiry into the Nature, Origin, and Fabrication of Life* (New York: Columbia University Press, 1991), 3.

4. For an English translation, see *Discourse on the Method* [1637], trans. John Veitch (La Salle, IL: Open Court Publishing Company, date unknown), 60–61, https://archive.org/details/discourseonthem00descuoft/page/60/mode/2up.

5. Rosen, *Life Itself*, 20.

6. Rosen, *Life Itself*, 21.

7. Roger Penrose, *Shadows of the Mind: A Search for the Missing Science of Consciousness* (New York: Oxford UP, 1994), 64.

8. A. W. Moore, *Gödel's Theorem: A Very Short Introduction* (Oxford: Oxford University Press, 2022), 1–12. The paper is "On Formally Undecidable Propositions of *Principia Mathematica* and Related Systems I." It was intended as a Part 1 with a sequel, Part 2, to follow (hence the "I" at the end) but it never did.

9. Robert Rosen, *Essays on Life Itself* (New York: Columbia University Press, 2000), v.

10. Rosen, *Essays on Life Itself*, 2.

11. This discovery would be revived in a more contemporary form by intelligent design advocates, such as the engineer Robert J. Marks in *Non-Computable You: What You Do That Artificial Intelligence Never Will* (Seattle, WA: Discovery Institute Press, 2022). See too the rigorous work of computer scientist Selmer Bringsjord of Rensselaer Polytechnic Institute, who was interviewed for the series *Science Uprising*: "Three Things AI Machines Won't Be Able to Achieve," Discovery Science, *YouTube*, November 11, 2022, video, 1:20:06, https://youtu.be/T0fItlQd3pE?si=f--iOhnLJELemx2X.

12. Rosen, *Essays on Life Itself*, 7.

13. Rosen, *Essays on Life Itself*, 30.

14. Lars Löfgren, "An Axiomatic Explanation of Complete Self-Reproduction," *Bulletin of Mathematical Biophysics* 30, no. 3 (September 1968): 424, https://cepa.info/fulltexts/1844.pdf.

15. René Thom, *Structural Stability and Morphogenesis* (Reading, MA: W. A. Benjamin, Inc., 1975), 1.

16. Thom, *Structural Stability and Morphogenesis*, 1.

17. Thom, *Structural Stability and Morphogenesis*, 2.

18. Eric H. Anderson, "A Factory That Builds Factories That Build Factories..." in *Evolution and Intelligent Design in a Nutshell* (Seattle, WA: Discovery Institute Press, 2020), 65–86.

19. "Demise of the Gene," *Evolution News and Science Today*, September 19, 2012, https://evolutionnews.org/2012/09/demise_of_the_g/.

20. Denis Noble, "It's Time to Admit That Genes Are Not the Blueprint for Life," *Nature* 626 (February 5, 2024): 254–255.

21. Ken Richardson, "It's the End of the Gene as We Know It," *Nautilus*, January 2, 2019, https://nautil.us/its-the-end-of-the-gene-as-we-know-it-237288/.

22. John A. Stamatoyannopoulos, "What Does Our Genome Encode?," *Genome Research* 22, no. 9 (September 2012): 1602–1611.

CHAPTER 8

1. Bill Gates, Nathan Myhrvold, and Peter Rinearson, *The Road Ahead* (New York: Viking, 1995), 188.

CHAPTER 9

1. Denis Noble, *The Music of Life: Biology Beyond Genes* (Oxford, UK: Oxford University Press, 2006).

CHAPTER 10

1. For an extensive recent literature roundup, see Richard Sternberg, Casey Luskin, and Jonathan McLatchie, "Here's a Far from Exhaustive (Yet Still Exhausting) List of Papers Discovering Function for 'Junk' DNA," *Evolution News and Science Today*, May 1, 2024, https://evolutionnews.org/2024/05/heres-a-far-from-exhaustive-yet-still-exhausting-list-of-papers-discovering-function-for-junk-dna/.
2. Explains Sternberg, "To say, as some do (e.g., Dan Graur), that at most 10 percent of human/mammalian DNA is functional is to say that 2 percent protein-coding DNA + 8 percent non-protein-coding DNA comprise that 10 percent, leaving the remaining 90 percent non-functional."
3. Jonathan Wells, "Glossary," in *The Myth of Junk DNA* (Seattle, WA: Discovery Institute Press, 2011), 161–169.
4. Rat Genome Sequencing Project Consortium, "Genome Sequence of the Brown Norway Rat Yields Insights into Mammalian Evolution," *Nature* 428 (April 1, 2004): 493–521.

CHAPTER 11

1. Jonathan Wells, "Glossary," in *The Myth of Junk DNA* (Seattle, WA: Discovery Institute Press, 2011), 161–169.
2. Robert Alicki, "Information Is Not Physical," *ArXiv*, February 11, 2014, https://arxiv.org/abs/1402.2414.
3. Richard Liangchen Wang, "Information Is Non-Physical: The Rules Connecting Representation and Meaning Do Not Obey the Laws of Physics," *Journal of Information Science* (December 21, 2022), https://journals.sagepub.com/doi/full/10.1177/01655515221141040.
4. Pentti Kanerva, "Hyperdimensional Computing: An Introduction to Computing in Distributed Representation with High-Dimensional Random Vectors," *Cognitive Computation* 1 (2009): 139–159.
5. Roger Penrose, "Précis of *The Emperor's New Mind: Concerning Computers, Minds, and the Laws of Physics*," *Behavioral and Brain Sciences* 13, no. 4 (December 1990): 643–705.

CHAPTER 12

1. F. H. C. Crick, "On Protein Synthesis," in *Symposia of the Society for Experimental Biology, XII: The Biological Replication of Macromolecules*, ed. F. K. Sanders (Cambridge, UK: Cambridge University Press, 1958), 153.

2. Crick, "On Protein Synthesis," 152.
3. Barbara McClintock, "The Significance of Responses of the Genome to Challenge" (lecture, *The Nobel Foundation,* Stockholm, Sweden, December 8, 1983). Barbara McClintock, "The Significance of Responses of the Genome to Challenge," *Science* 226, no. 4676 (November 26, 1984): 792–801.
4. J. A. Shapiro, "Natural Genetic Engineering in Evolution," *Genetica* 86 (1992): 99–111.
5. Aaron David Goldman and Laura F. Landweber, "What Is a Genome?," *PLOS Genetics* 12, no. 7 (2016): e1006181.

CHAPTER 13

1. A. N. Wilson, *God's Funeral: The Decline of Faith in Western Civilization* (New York: W. W. Norton, 1999), xi.
2. Jonathan Wells, "Glossary," in *The Myth of Junk DNA* (Seattle, WA: Discovery Institute Press, 2011), 168.
3. Seth Lloyd, "Ultimate Physical Limits to Computation," *Nature* 406 (2000): 1047–1054.

CHAPTER 14

1. David P. Goldman, "The Soloveitchik Solution," *Tablet,* May 10, 2023, https://www.tabletmag.com/sections/israel-middle-east/articles/rav-soloveitchik-solution.
2. *Kabbalah* in Hebrew means "transmission," "acceptance," "tradition." In the broadest sense it refers to the tradition, handed down through generations, that explains the often unclear or self-contradictory scriptural text in its original languages.

    In contrast with the idea of *sola scriptura*, that reading scripture alone is sufficient and valid, proponents of tradition, whether Jewish, Catholic, or Orthodox Christian, point out that the biblical text, especially the Old Testament, comes down to us without vowels, punctuation, or other clarifications as to what it's getting at. Imagine trying to make sense of this book if we left the vowels and periods out. The translation of the King James Bible, as one example, could not have been done without a tradition, a *kabbalah*.

    Traditions for understanding the Bible appear historically in versions ranging from the straightforward to the difficult or obscure. In the narrower sense, *kabbalah* refers to these more obscure or esoteric veins, often but not always with a mystical flavor, later recorded in works such as the *Zohar*.

    Many have noted that whether Jewish, Christian, or otherwise, world religious traditions tend to converge on various themes. C. S. Lewis called this convergence the Tao (now usually rendered as Dao). On one of our first meetings, Richard Sternberg phrased it differently, as the Great Tradition (not his coining).

    An early reader of this book raised the concern that, according to his understanding, there is something occult about the *kabbalah*. But this is to mistake misuse for use. We can't even begin to read the Bible without a *kabbalah*. Rabbi Soloveitchik, whom I cite in this chapter, is about as mainstream a modern religious figure as you could ask for. It's true that certain occult flim-flam artists have co-opted some symbols from Jewish *kabbalistic* tradition. Similarly, some writers and teachers have co-opted Christian symbols and sought to put them to occult purposes. The work of psychologist Carl Jung, arguably, provides an example. As I was writing this book, a man approached me at Trader Joe's wishing to acquaint me with theosophy, another such attempt which he described as "advanced Christianity." But such things are the occultists' problem, not ours.

3. Goldman, "The Soloveitchik Solution."

4. Goldman, "The Soloveitchik Solution."

5. Meir Triebitz, "Rabbi Joseph B. Soloveitchik's Lectures on Genesis, VI Through IX," *Hakirah* 29 (2021): 35–36n38, https://hakirah.org/Vol29Triebitz.pdf.

6. Goldman, "The Soloveitchik Solution." In email correspondence, Goldman noted that the final word in the original involved a typo—"finitude" instead of the intended "infinitude." That has been corrected here.

7. Charles Darwin, *Hypothesis of Pangenesis* [1865]. See *The Complete Work of Charles Darwin Online*, ed. John van Wyhe, https://darwin-online.org.uk/content /frameset?itemID=CUL-DAR51.C36-C74&keywords=pangenesis& viewtype=text&pageseq=1.

8. D. L. Nanney, "Epigenetic Control Systems," *PNAS* 44, no. 7 (July 15, 1958): 712–717, https://www.pnas.org/doi/abs/10.1073/pnas.44.7.712.

9. The comment is on Exodus 3:2. Rabbi Joseph B. Soloveitchik, *Chumash Mesoras Harav: Sefer Shemos* [Book of Exodus] (New York: Orthodox Union Press, 2014), 22.

10. The inference appears in *Yoma* 21a, *Megillah* 10b, and *Bava Batra* 99a. See Shlomo Chaim Kesselman, "The Ark of the Covenant," *Chabad.org*, accessed January 21, 2025, https://www.chabad.org/library/article_cdo/aid/4277479/jewish/The-Ark -of-the-Covenant.htm.

CHAPTER 15

1. Cited by D. A. Rees, "Platonism and the Platonic Tradition," *The Encyclopedia of Philosophy* (New York: Macmillan & Free Press, 1967), 6: 340.

2. Daniel Witt, "Science versus the Oldest Anti-Intelligent Design Argument," *Evolution News and Science Today*, May 13, 2024, https://evolutionnews.org/2024 /05/science-versus-the-oldest-anti-intelligent-design-argument/.

3. Witt, "Science versus the Oldest Anti-Intelligent Design Argument."

4. Andrew Gregory, introduction to and notes on *Timaeus and Critias*, by Plato, trans. Robin Waterfield (Oxford, UK: Oxford University Press, 2008), xiii.

5. Gregory, introduction, *Timaeus*, xiv.

6. Gregory, introduction, *Timaeus*, xvii.

7. Gregory, introduction, *Timaeus*, xvii.

8. Gregory, introduction, *Timaeus*, xx.

9. Gregory, introduction, *Timaeus*, xxi.

10. Gregory, introduction, *Timaeus*, lii.

11. It's cross-cultural, too: the Jewish *Zohar* commenting on the first verse of Genesis (*Terumah* 161b), from elsewhere in the ancient world, likewise has the Creator looking in a source of preexisting wisdom or information, the book of the Torah, and copying from it (not to be taken literally, I'm sure).

12. Plato, *Timaeus*, 17.

13. Gregory, *Timaeus*, 128.

14. Michael Denton, *Nature's Destiny: How the Laws of Biology Reveal Purpose in the Universe* (New York: The Free Press, 1998). The five books in his Privileged Species series are *Fire-Maker* (2016) *The Wonder of Water* (2017), *Children of Light* (2018), *The Miracle of the Cell* (2020), and *The Miracle of Man* (2022), all from Discovery Institute Press.

15. Michael Denton, "Supreme Elegance: The Fine-Tuning of the Properties of Matter for Life on Earth," *Evolution News and Science Today*, August 6, 2024, https://evolutionnews.org/2024/08/supreme-elegance-the-fine-tuning-of-the-properties-of-matter-for-life-on-earth/. Emphasis in original.

16. Charles Townes, quoted in Bonnie Azab Powell, "'Explore as Much as We Can': Nobel Prize Winner Charles Townes on Evolution, Intelligent Design, and the Meaning of Life," *UC Berkeley News*, June 17, 2005, https://newsarchive.berkeley.edu/news/media/releases/2005/06/17_townes.shtml.

17. Plato, *Timaeus*, 38.

18. Plato, *Timaeus*, 67–68.

19. Stuart Burgess, "Universal Optimal Design in the Vertebrate Limb Pattern and Lessons for Bioinspired Design," *Bioinspiration & Biomimetics* 19, no. 5 (2024), https://iopscience.iop.org/article/10.1088/1748-3190/ad66a3.

20. Richard Sternberg, "*Logos* and Materialism: Why Aristotle Favors Intelligent Design and Not Physicalist Thomism," in Ann Gauger, ed., *God's Grandeur: The Catholic Case for Intelligent Design* (Manchester, NH: Sophia Institute Press, 2023), 269–289.

21. Sternberg, "*Logos* and Materialism," 280.

22. Sternberg, "*Logos* and Materialism," 285.

23. Sternberg, "*Logos* and Materialism," 276.

24. Sternberg, "*Logos* and Materialism," 284.

CHAPTER 16

1. Stephen Meyer, *Return of the God Hypothesis: Three Scientific Discoveries That Reveal the Mind Behind the Universe* (New York: HarperOne, 2021), 182–188.

2. J. G. M. Thewissen and Sunil Bajpai, "Whale Origins as a Poster Child for Macroevolution: Fossils Collected in the Last Decade Document the Ways in which Cetacea (Whales, Dolphins, and Porpoises) Became Aquatic, a Transition That Is One of the Best Documented Examples of Macroevolution in Mammals," *BioScience* 51, no. 12 (December 2001): 1037–1049.

3. Charles Darwin, *On the Origin of Species by Means of Natural Selection* (London: John Murray, 1859), 184.

4. For a brief overview of the responses Darwin received that led him to remove the bear-to-whale evolution passage from subsequent editions of *The Origin of Species*, see Robert F. Shedinger, *Darwin's Bluff: The Mystery of the Book Darwin Never Finished* (Seattle, WA: Discovery Institute Press, 2024), 130–131.

5. Thewissen and Bajpai, "Whale Origins as a Poster Child for Macroevolution," 1037–1049.

6. Jonathan Wells, "From Bears to Whales: A Difficult Transition," *Evolution News and Science Today*, July 17, 2018, https://evolutionnews.org/2018/07/from-bears-to -whales-a-difficult-transition/.

7. Sternberg discussed these things in a lengthy interview for the series *Science Uprising*, which I draw on here. "National Geographic View of Fossil Record Is WRONG Says Evolutionary Biologist Richard Sternberg," Discovery Science, *YouTube*, December 14, 2021, video, 54:28, https://youtu.be/glgXFGW _K6g?si=M9f_pfYMqaU35e6T.

8. Emilie Le Beau Lucchesi, "How Many Whales Are Left in the World?," *Discover*, October 26, 2023, https://www.discovermagazine.com/planet-earth/how-many -whales-are-left-in-the-world.

9. Stuart Dredge, "Tesla Founder Elon Musk Buys James Bond's Lotus Esprit Submarine Car," *Guardian*, October 18, 2013, https://www.theguardian.com /technology/2013/oct/18/tesla-elon-musk-james-bond-lotus-submarine-car.

10. Rick Durrett and Deena Schmidt, "Waiting for Two Mutations: With Applications to Regulatory Sequence Evolution and the Limits of Darwinian Evolution," *Genetics* 180, no. 3 (November 2008): 1501–1509.

11. Hugo de Vries, *Species and Varieties: Their Origin by Mutation*, 2nd ed. (Chicago: Open Court, 1904), 825–826.

12. Meyer, *Return of the God Hypothesis*, 24. See also Ted Davis, "The Faith of a Great Scientist: Robert Boyle's Religious Life, Attitudes, and Vocation," and in particular the section titled "Did God Have Any Choice When He Made the World?" *BioLogos*, August 8, 2013, https://biologos.org/articles/ the-faith-of-a-great-scientist-robert-boyles-religious-life-attitudes-and-vocation.

13. Michael Egnor, "Why Aristotle and Aquinas?," *Evolution News and Science Today*, July 25, 2017, https://evolutionnews.org/2017/07/why-aristotle-and-aquinas / https://evolutionnews.org/2017/07/why-aristotle-and-aquinas/.

14. Michael Denton, "Aristotle Rediscovered: What Exactly Is the 'Mechanism' for Intelligent Design?," *Evolution News and Science Today*, April 21, 2015, https:// evolutionnews.org/2015/04/aristotle_redis/.

Chapter 17

1. "Richard Dawkins vs Ayaan Hirsi Ali: The God Debate," *UnHerd*, June 3, 2024, https://unherd.com/watch-listen/the-god-debate/.

2. The phrase is from Eugene Wigner, "The Unreasonable Effectiveness of Mathematics in the Natural Sciences," *Communications on Pure and Applied Mathematics* 13, no. 1 (May 11, 1959): 1–14.

3. See, for example, Stephen Meyer, "The Demarcation of Science and Religion," *Discovery Institute*, January 1, 2000, https://www.discovery.org/a/3524/.

4. One sees strong hints of this bias for science focused on the here and now in National Academy of Sciences member Philip Skell's commentary on the results of

an informal survey he conducted and reported on in the journal *The Scientist* ("Why Do We Invoke Darwin?," August 28, 2005). Skell writes:

> My own research with antibiotics during World War II received no guidance from insights provided by Darwinian evolution. Nor did Alexander Fleming's discovery of bacterial inhibition by penicillin. I recently asked more than 70 eminent researchers if they would have done their work differently if they had thought Darwin's theory was wrong. The responses were all the same: No.
>
> I also examined the outstanding biodiscoveries of the past century: the discovery of the double helix; the characterization of the ribosome; the mapping of genomes; research on medications and drug reactions; improvements in food production and sanitation; the development of new surgeries; and others. I even queried biologists working in areas where one would expect the Darwinian paradigm to have most benefited research, such as the emergence of resistance to antibiotics and pesticides. Here, as elsewhere, I found that Darwin's theory had provided no discernible guidance, but was brought in, after the breakthroughs, as an interesting narrative gloss.
>
> In the peer-reviewed literature, the word "evolution" often occurs as a sort of coda to academic papers in experimental biology. Is the term integral or superfluous to the substance of these papers? To find out, I substituted for "evolution" some other word—"Buddhism," "Aztec cosmology," or even "creationism." I found that the substitution never touched the paper's core. This did not surprise me. From my conversations with leading researchers it had become clear that modern experimental biology gains its strength from the availability of new instruments and methodologies, not from an immersion in historical biology.

5. If Sternberg is wrong, then of course that takes nothing away from other arguments for intelligent design.

6. Stephen Meyer emphasizes that the ID explanation for the Cambrian explosion, unlike every purely materialistic theory for the event, posits a presently acting type of cause active in the present with the demonstrated capacity to generate novel information (namely, a designing intelligence), a capacity that is a prerequisite for generating the many novel body plans of the Cambrian explosion.

7. Casey Luskin, "2024 Nobel Prize Awarded for the Discovery of Function for a Type of 'Junk DNA,'" *Evolution News and Science Today*, December 17, 2024, https://evolutionnews.org/2024/12/2024-nobel-prize-awarded-for-the -discovery-of-function-for-a-type-of-junk-dna/.

8. "Fossil Explosions in the History of Life: Paleontologist Günter Bechly," Discovery Science, *YouTube*, December 16, 2024, video, 51:47, https://youtu.be /izzNeLFTyKU?si=3v7NCGdZXQiRQUKX.

9. Stephen C. Meyer and Günter Bechly, "The Fossil Record and Universal Common Ancestry," in *Theistic Evolution: A Scientific, Philosophical, and Theological Critique*, ed. J. P. Moreland et al. (Wheaton: Crossway, 2017), 353.

10. Meyer and Bechly, "The Fossil Record and Universal Common Ancestry," 331–362.

11. Stephen Meyer, *Return of the God Hypothesis: Three Scientific Discoveries That Reveal the Mind Behind the Universe* (New York: HarperOne, 2021), 415.

12. Meyer, *Return of the God Hypothesis*, 415.

EPILOGUE

1. I shared the results with my children and suggested that they look into the history of the North African Barbary pirates, who took many a European or American hostage over the centuries, including women who were ravished or inducted into harem life. An entire Irish coastal village was abducted in 1631 by these pirates. Only two residents returned. President Jefferson fought our first foreign war, "to the shores of Tripoli," over rescuing pirate hostages. Mozart wrote an opera on the theme, *The Abduction from the Seraglio*, as did Rossini, *The Italian Girl in Algiers*. About my most recent Native American ancestor, the website *FamilySearch.org* nails this down with a startling certainty to a daughter of the Seneca leader Tanacharison (1700–1754), the so-called Half King, who helped start the French and Indian War. She married an Irishman named Owens.

2. Roger Lockhurst, introduction to *Late Victorian Gothic Tales* (Oxford, UK: Oxford University Press, 2015), xx.

3. Lockhurst, *Late Victorian Gothic Tales*, xv.

4. Quoted in Jacques Monod, *Chance & Necessity* (New York: Vintage Books, 1972), v.

5. Monod, *Chance & Necessity*, 180.

6. Mike Cernovich (@Cernovich), "You are all sickly and unhealthy. REVOLTING! I am 47 and more vital. This is pathetic. The old men are supposed to fear the young," *X*, November 8, 2024, https://x.com/Cernovich/status /1854948214694461882.

7. Alfred Russel Wallace, quoted in Harold Begbie, "New Thoughts on Evolution," *Daily Chronicle*, November 3 and 4, 1910, 4, https://people.wku.edu/charles.smith /wallace/S746.htm.

8. Wallace, interview with Begbie, *Daily Chronicle*.

INDEX

www.ingramcontent.com/pod-product-compliance
Lightning Source LLC
Chambersburg PA
CBHW020156200326
41521CB00006B/395